Leap Motion Development Essentials

Leverage the power of Leap Motion to develop
and deploy a fully interactive application

Mischa Spiegelmock

PUBLISHING

BIRMINGHAM - MUMBAI

Leap Motion Development Essentials

First published: October 2013

Production Reference: 1211013

Published by Packt Publishing Ltd.
Livery Place
35 Livery Street
Birmingham B3 2PB, UK.

ISBN 978-1-84969-772-9

www.packtpub.com

Cover Image by Suresh Mogre (suresh.mogre.99@gmail.com)

Credits

Author
Mischa Spiegelmock

Reviewers
Shaun Walker
Carlos Zinato

Acquisition Editors
Erol Staveley
Rebecca Youe

Commissioning Editor
Mohammed Fahad

Technical Editors
Vrinda Nitesh Bhosale
Pratik More
Amit Ramadas

Project Coordinator
Suraj Bist

Proofreader
Bernadette Watkins

Indexer
Hemangini Bari

Production Coordinator
Manu Joseph

Cover Work
Manu Joseph

About the Author

Mischa Spiegelmock is an accomplished software engineer from the San Francisco Bay Area. Slightly infamous from light-hearted technical pranks from his youth, he is now a respectable CTO at a healthcare software startup. His passions are architecting elegant and useful programs and sharing his insights into software design with others in a straightforward and entertaining fashion.

This book would not exist without the extremely passionate people at Leap Motion, whose vision and heroic efforts to improve the interactions between humans and computers should be commended. Special mention to Elizabeth Ruscitto for taking the time to provide me with informative answers to my queries. I would also like to thank Hep Svadja Hepic Photography, for the amazing pictures.

About the Reviewers

Shaun Walker is a software and web developer from Tasmania, Australia. He enjoys tinkering with electronics, listening to music, playing computer games, and working on small projects that generally never get finished.

Shaun has a Bachelor's degree in Computing, majoring in Human Interface Technology, part of the HITLab, an International collaboration effort between a few universities around the world. His areas of focus during his studies were in augmented reality, virtual reality, and gesture-based interactions.

With a history in corporate IT support, working on network integration, software and web development, his main employment is with a Brisbane-based medical start-up.

Ramblings by Shaun can be found at his blog at `http://theshaun.com` or on Twitter `@theshaun`.

Carlos Zinato, aged 25, is from São Paulo, Brazil. Currently, he's living in Miami Beach, FL, developing mobile apps and games, playing bass guitar, and getting to know people and places.

Working with tools such as Objective-C, Leap Motion, Titanium, and Unity3D, Carlos makes mobile apps and games and this is what he loves to do.

Carlos is currently aiming to keep making awesome apps and games with skilled friends and teams. He not only cares about the money but, more importantly, about the knowledge. Carlos wants to keep working with brilliant people that enhance his job, his life, and his pocket.

www.packtpub.com

Support files, eBooks, discount offers and more

You might want to visit www.packtpub.com for support files and downloads related to your book.

Did you know that Packt offers eBook versions of every book published, with PDF and ePub files available? You can upgrade to the eBook version at www.packtpub.com and as a print book customer, you are entitled to a discount on the eBook copy. Get in touch with us at service@packtpub.com for more details.

At www.packtpub.com, you can also read a collection of free technical articles, sign up for a range of free newsletters and receive exclusive discounts and offers on Packt books and eBooks.

www.packtpub.com

Do you need instant solutions to your IT questions? PacktLib is Packt's online digital book library. Here, you can access, read and search across Packt's entire library of books.

Why Subscribe?

- Fully searchable across every book published by Packt
- Copy and paste, print and bookmark content
- On demand and accessible via web browser

Free Access for Packt account holders

If you have an account with Packt at www.packtpub.com, you can use this to access PacktLib today and view nine entirely free books. Simply use your login credentials for immediate access.

About the Reviewers

Shaun Walker is a software and web developer from Tasmania, Australia. He enjoys tinkering with electronics, listening to music, playing computer games, and working on small projects that generally never get finished.

Shaun has a Bachelor's degree in Computing, majoring in Human Interface Technology, part of the HITLab, an International collaboration effort between a few universities around the world. His areas of focus during his studies were in augmented reality, virtual reality, and gesture-based interactions.

With a history in corporate IT support, working on network integration, software and web development, his main employment is with a Brisbane-based medical start-up.

Ramblings by Shaun can be found at his blog at `http://theshaun.com` or on Twitter `@theshaun`.

Carlos Zinato, aged 25, is from São Paulo, Brazil. Currently, he's living in Miami Beach, FL, developing mobile apps and games, playing bass guitar, and getting to know people and places.

Working with tools such as Objective-C, Leap Motion, Titanium, and Unity3D, Carlos makes mobile apps and games and this is what he loves to do.

Carlos is currently aiming to keep making awesome apps and games with skilled friends and teams. He not only cares about the money but, more importantly, about the knowledge. Carlos wants to keep working with brilliant people that enhance his job, his life, and his pocket.

www.packtpub.com

Support files, eBooks, discount offers and more

You might want to visit www.packtpub.com for support files and downloads related to your book.

Did you know that Packt offers eBook versions of every book published, with PDF and ePub files available? You can upgrade to the eBook version at www.packtpub.com and as a print book customer, you are entitled to a discount on the eBook copy. Get in touch with us at service@packtpub.com for more details.

At www.packtpub.com, you can also read a collection of free technical articles, sign up for a range of free newsletters and receive exclusive discounts and offers on Packt books and eBooks.

www.packtpub.com

Do you need instant solutions to your IT questions? PacktLib is Packt's online digital book library. Here, you can access, read and search across Packt's entire library of books.

Why Subscribe?

- Fully searchable across every book published by Packt
- Copy and paste, print and bookmark content
- On demand and accessible via web browser

Free Access for Packt account holders

If you have an account with Packt at www.packtpub.com, you can use this to access PacktLib today and view nine entirely free books. Simply use your login credentials for immediate access.

Table of Contents

Preface

Once in a great while, a revolutionary and exciting new way of interacting with technology comes along. Leap Motion makes natural gesture-based interfaces a reality, and gives software developers access to a large and powerful set of features and capabilities. This guide is for software engineers who wish to get an overview of the Leap Software Development Kit modules, types and interfaces in C++ along with some guidelines for getting the most out of your Leap device and creating usable, gestural software interfaces.

What this book covers

Chapter 1, *Leap Motion SDK – A Quick Start*, explores how to begin using the Leap C++ SDK right away, with a sample mouse control program. It covers how to receive frame updates and read finger position data.

Chapter 2, *Real Talk – Real Time*, guides you through writing a multithreaded MIDI controller, which uses a blocking OS call without sacrificing responsiveness.

Chapter 3, *Actual Gestures*, covers a high-level discussion of gesture interfaces and a look at the available Leap SDK recognizers. It also covers creating an interface to manipulate Windows OS.

Chapter 4, *Leap and the Web*, teaches you how to create Leap-enabled web pages, using JavaScript with the LeapJS library, with no additional installation or configuration.

Chapter 5, *HTML5 Antics in 3D*, combines LeapJS, WebGL, and Three.js together to manipulate objects in 3D space in a web page using Leap Motion.

What you need for this book

You will need a working knowledge of basic C++ and a compiler. Additional demonstration operating system-specific code is provided for Mac OS X APIs, but neither OS X nor familiarity with it is a requirement. An understanding of basic geometry and concepts such as vectors is useful for spatial manipulations.

Who this book is for

This book is for developers with an interest in using the Leap Motion input device with their software. This book gives a broad overview of most of the available functionality in the Leap SDK, which can be transferred to any of the supported language interfaces, as the data types and routines are nearly identical.

Conventions

In this book, you will find a number of styles of text that distinguish between different kinds of information. Here are some examples of these styles, and an explanation of their meaning.

Code words in text are shown as follows: "We can include other contexts through the use of the `include` directive."

A block of code is set as follows:

```
if (frame.hands().empty()) return;

const Leap::Hand firstHand = frame.hands()[0];
const Leap::FingerList fingers = firstHand.fingers();
```

When we wish to draw your attention to a particular part of a code block, the relevant lines or items are set in bold:

```
namespace leapmidi {

typedef double midi_control_value_raw;
typedef unsigned short midi_control_value;
```

New terms and **important words** are shown in bold. Words that you see on the screen, in menus or dialog boxes, for example, appear in the text like this: "Prepare to be astounded when you point at the middle of your screen and the transfixing message, **You are pointing at (0.519522, 0.483496, 0)**, is revealed".

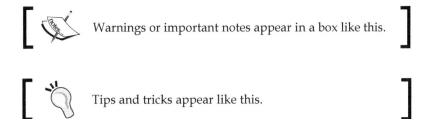

Warnings or important notes appear in a box like this.

Tips and tricks appear like this.

Reader feedback

Feedback from our readers is always welcome. Let us know what you think about this book — what you liked or may have disliked. Reader feedback is important for us to develop titles that you really get the most out of.

To send us general feedback, simply send an e-mail to feedback@packtpub.com, and mention the book title via the subject of your message.

If there is a topic that you have expertise in and you are interested in either writing or contributing to a book, see our author guide on www.packtpub.com/authors.

Customer support

Now that you are the proud owner of a Packt book, we have a number of things to help you to get the most from your purchase.

Downloading the example code

You can download the example code files for all Packt books you have purchased from your account at http://www.packtpub.com. If you purchased this book elsewhere, you can visit http://www.packtpub.com/support and register to have the files e-mailed directly to you. You can also find the updated code samples at https://github.com/openleap/leapbook.

Errata

Although we have taken every care to ensure the accuracy of our content, mistakes do happen. If you find a mistake in one of our books—maybe a mistake in the text or the code—we would be grateful if you would report this to us. By doing so, you can save other readers from frustration and help us improve subsequent versions of this book. If you find any errata, please report them by visiting `http://www.packtpub.com/submit-errata`, selecting your book, clicking on the **errata submission form** link, and entering the details of your errata. Once your errata are verified, your submission will be accepted and the errata will be uploaded on our website, or added to any list of existing errata, under the Errata section of that title. Any existing errata can be viewed by selecting your title from `http://www.packtpub.com/support`.

Piracy

Piracy of copyright material on the Internet is an ongoing problem across all media. At Packt, we take the protection of our copyright and licenses very seriously. If you come across any illegal copies of our works, in any form, on the Internet, please provide us with the location address or website name immediately so that we can pursue a remedy.

Please contact us at `copyright@packtpub.com` with a link to the suspected pirated material.

We appreciate your help in protecting our authors, and our ability to bring you valuable content.

Questions

You can contact us at `questions@packtpub.com` if you are having a problem with any aspect of the book, and we will do our best to address it.

Leap Motion SDK – A Quick Start

The Leap Motion is a peripheral input device that allows users to interact with the software through hand gestures. This chapter will explain the features and software interface exposed by the Leap Motion SDK, which will allow us to take advantage of the hand motion input.

An overview of the SDK

The Leap device uses a pair of cameras and an infrared pattern projected by LEDs to generate an image of your hands with depth information. A very small amount of processing is done on the device itself, in order to keep the cost of the units low.

The images are post-processed on your computer to remove noise, and to construct a model of your hands, fingers, and pointy tools that you are holding.

As an application developer, you can make use of this data via the Leap software developer kit, which contains a powerful high-level API for easily integrating gesture input into your applications. Because developers do not want to go to, the trouble of processing raw input in the form of depth-mapped images skeleton models and point cloud data, the SDK provides abstracted models that report what your user is doing with their hands. With the SDK you can write applications that make use of some familiar concepts:

- All hands detected in a frame, including rotation, position, velocity, and movement since an earlier frame
- All fingers and pointy tools (collectively known as "pointables") recognized as attached to each hand, with rotation, position, and velocity
- The exact pixel location on a display pointed at by a finger or tool
- Basic recognition of gestures such as swipes and taps
- Detection of position and orientation changes between frames

Quick start

Congratulations on your purchase of a genuine fine quality Leap Motion gesture input device! This handy guide will walk you through the assembly, proper usage, and care of your Leap Motion.

To get started, remove your Leap Motion SDK and Leap Motion™ device from the box and unpack the shared object files and headers from their shrink-wrap. Gently place your SDK in a handy directory and fire up your favorite IDE to begin.

Downloading the example code

You can download the example code files for all Packt books you have purchased from your account at http://www.packtpub.com. If you purchased this book elsewhere, you can visit http://www.packtpub.com/support and register to have the files e-mailed directly to you.

We'll get things going right away with a short C++ application to illustrate how to interact with the Leap SDK to receive events and input.

```cpp
#include <iostream>
#include <Leap.h>

class Quickstart : public Leap::Listener {
public:
    virtual void onConnect(const Leap::Controller &);
    virtual void onFrame(const Leap::Controller &);
};
```

To interact with the Leap software, we will begin by creating a subclass of `Leap::Listener` and defining the callback methods we wish to receive. While it is possible to poll the controller for the current frame, generally you will want to make your program as responsive as possible which is most easily accomplished by acting on input events immediately via callbacks. In case you're wondering, the primary available callback methods are:

- `onInit`: This indicates that the listener is added to a controller and called only once.

- `onExit`: This indicates that the controller is destroyed or the listener is removed.

- `onConnect`: This indicates that the device is connected and recognized by the driver and is ready to start processing frames.

- `onDisconnect`: This indicates that the device is disconnected. Connection state can also be polled by checking `controller.isConnected()` so that you don't need to keep track of whether the controller is plugged in or not. (An earlier version of the SDK lacked this state accessor but the kind folks at Leap Motion realized us devs are really lazy).
- `onFrame`: This indicates that a new frame of input data has been captured and processed. This is the only handler you really need to implement if you want to make use of the Leap.

Let's implement our `onConnect` handler real quick so that we can verify that the controller driver and SDK are communicating with the device properly. If everything is working as it should be, the following code should cause a message to be emitted on stdout when the device is plugged in to a USB port:

```cpp
void Quickstart::onConnect(const Leap::Controller &controller) {
    std::cout << "Hello, Leap user!\n";
}
```

We can display a friendly contrived greeting to ourselves when the program is run with the controller connected and the driver software is running. This will cause breathless anticipation in your users as they prepare themselves to experience the magic and wonder of this fantastic new input technology.

A listener is attached to a `Leap::Controller`, which acts as the primary interface between the Leap driver and your application. A controller tracks processed frames, device connection state, configuration parameters, and invokes callback methods on a listener.

To begin receiving events, instantiate a new listener and a controller:

```cpp
int main() {
    // create instance of our Listener subclass
    Quickstart listener;

    // create generic Controller to interface with the Leap device
    Leap::Controller controller;

    // tell the Controller to start sending events to our Listener
    controller.addListener(listener);
    ...
```

If you place your hands over the device `Quickstart::onFrame()` will start being called. Let's create an `onFrame` handler that reports on the horizontal velocity of the first finger or tool detected:

```cpp
void Quickstart::onFrame(const Leap::Controller &controller) {
    const Leap::Frame frame = controller.frame();
```

`controller.frame()` returns a `Frame` instance, which contains information detected about our scene at a specific point in time. It has a single optional parameter `history`, which allows you to travel backwards through the misty sands of time to compare frames and determine hand changes over time. Unfortunately this time machine is rather limited; only about 60 previous frames are stored in the controller.

```
// do nothing unless hands are detected
if (frame.hands().empty()) return;
```

Get used to making these sorts of checks. You'll be seeing a lot more of them. Here, we have no interest in processing this frame unless there are hands in it.

```
// first detected hand
const Leap::Hand firstHand = frame.hands()[0];
// first pointable object (finger or tool)
const Leap::PointableList pointables = firstHand.pointables();
if (pointables.empty()) return;
const Leap::Pointable firstPointable = pointables[0];
```

All fingers attached to a hand and all tools that the hand is grasping are returned as a **PointableList**, which behaves like an `std::vector`, including providing an iterator for people who are into that. Most commonly we will want to find out where a pointable is in space and how fast it is moving, which we can easily find out with `tipPosition()` and `tipVelocity()` respectively. Both return `Leap::Vector`s consisting of X, Y, and Z components.

```
std::cout << "Pointable X velocity: " << firstPointable.
tipVelocity()[0] << endl;
```

If you wave an outstretched finger or a tool (a chopstick works pretty well if you happen to have one lying around) back and forth over the controller you will be rewarded with the following riveting output:

```
Pointable X velocity: -223.937

Pointable X velocity: -117.421

Pointable X velocity: -242.293

Pointable X velocity: -141.43

Pointable X velocity: -61.9314

Pointable X velocity: 9.85328

Pointable X velocity: 41.9575

Pointable X velocity: 71.7436

Pointable X velocity: 96.0459

Pointable X velocity: 116.465
```

Leftwards motion is represented by negative values (mm/s). Rightwards motion is positive.

A note on the sample code

Because of frequent changes to the SDK, your best bet for finding the most up-to-date code samples is to check out the GitHub repository.

All sample code can be found at `https://github.com/openleap/ leapbook`. This program and others like it can be built using the following command on Mac OS X or Linux using GCC or clang:

```
$ g++ quickstart.cpp -lLeap -Lpath/to/Leap_SDK/lib/
libc++ -Ipath/to/Leap_SDK/include -o quickstart
```

Note that `path/to/Leap_SDK` should be replaced with the location of your Leap_SDK directory. It may be helpful to set an environment variable with the path or install the libraries and headers system-wide.

Major SDK components

Now that we've written our first gesture-enabled program, let's talk about the major components of the Leap SDK. We'll visit each of these in more depth as we continue our journey.

Controller

The `Leap::Controller` class is a liaison between the controller and your code. Whenever you wish to do anything at all with the device you must first go through your controller. From a controller instance we can interact with the device configuration, detected displays, current and past frames, and set up event handling with our listener subclass.

Config

An instance of the `Config` class can be obtained from a controller. It provides a key/value interface to modify the operation of the Leap device and driver behavior. Some of the options available are:

- **Robust mode**: Somewhat slower frame processing but works better with less light.

- **Low resource mode**: Less accurate and responsive tracking, but uses less CPU and USB bandwidth.

- **Tracking priority**: Can prioritize either precision of tracking data or the rate at which data is sampled (resulting in approximately 4x data frame-rate boost), or a balance between the two (approximately 2x faster than the precise mode).
- **Flip tracking**: Allows you to use the controller with the USB cable coming out of either side. This setting simply flips the positive and negative coordinates on the X-axis.

Screen

A controller may have one or more `calibratedScreens`, which are computer displays in the field of view of the controller, which have a known position and dimensions. Given a pointable direction and a screen we can determine what the user is pointing at.

Math

Several math-related functions and types such as `Leap::Vector`, `Leap::Matrix`, and `Leap::FloatArray` are provided by `LeapMath.h`. All points in space, screen coordinates, directions, and normal are returned by the API as three-element vectors representing X, Y, and Z coordinates or unit vectors.

Frame

The real juicy information is stored inside each `Frame`. A `Frame` instance represents a point in time in which the driver was able to generate an updated view of its world and detect where screens, your hands, and pointables are.

Hand

At present the only body parts you can use with the controller are your hands. Given a frame instance we can inspect the number of hands in the frame, their position and rotation, normal vectors, and gestures. The hand motion API allows you to compare two frames and determine if the user has performed a translation, rotation, or scaling gesture with their hands in that time interval. The methods we can call to check for these interactions are:

- `Leap::Hand::translation(sinceFrame)`: Translation (also known as movement) returned as a `Leap::Vector` including the direction of the movement of the hand and the distance travelled in millimeters.

- `Leap::Hand::rotationMatrix(sinceFrame)`, `::rotationAxis(sinceFrame)`, `::rotationAngle(sinceFrame, axisVector)`: Hand rotation, either described as a rotation matrix, vector around an axis or float angle around a vector between –π and π radians (that's -180° to 180° for those of you who are a little rusty with your trigonometry).

- `Leap::Hand::scaleFactor(sinceFrame)`: Scaling represents the distance between two hands. If the hands are closer together in the current frame compared to `sinceFrame`, the return value will be less than 1.0 but greater than 0.0. If the hands are further apart the return value will be greater than 1.0 to indicate the factor by which the distance has increased.

Pointable

A `Hand` also can contain information about `Pointable` objects that were recognized in the frame as being attached to the hand. A distinction is made between the two different subclasses of pointable objects, `Tool`, which can be any slender, long object such as a chopstick or a pencil, and `Finger`, whose meaning should be apparent. You can request either fingers or tools from a `Hand`, or a list of pointables to get both if you don't care.

Finger positioning

Suppose we want to know where a user's fingertips are in space. Here's a short snippet of code to output the spatial coordinates of the tips of the fingers on a hand that is being tracked by the controller:

```
if (frame.hands().empty()) return;

const Leap::Hand firstHand = frame.hands()[0];
const Leap::FingerList fingers = firstHand.fingers();
```

Here we obtain a list of the fingers on the first hand of the frame. For an enjoyable diversion let's output the locations of the fingertips on the hand, given in the Leap coordinate system:

```
for (int i = 0; i < fingers.count(); i++) {
    const Leap::Finger finger = fingers[i];

    std::cout << "Detected finger " << i << " at position (" <<
        finger.tipPosition().x << ", " <<
        finger.tipPosition().y << ", " <<
```

```
            finger.tipPosition().z << ")" << std::endl;
    }
```

This demonstrates how to get the position of the fingertips of the first hand that is recognized in the current frame. If you hold three fingers out the following dazzling output is printed:

```
Detected finger 0 at position (-119.867, 213.155, -65.763)

Detected finger 1 at position (-90.5347, 208.877, -61.1673)

Detected finger 2 at position (-142.919, 211.565, -48.6942)
```

While this is clearly totally awesome, the exact meaning of these numbers may not be immediately apparent. For points in space returned by the SDK the Leap coordinate system is used. Much like our forefathers believed the Earth to be the cornerstone of our solar system, your Leap device has similar notions of centricity. It measures locations by their distance from the Leap origin, a point centered on the top of the device. Negative X values represent a point in space to the left of the device, positive values are to the right. The Z coordinates work in much the same way, with positive values extending towards the user and negative values in the direction of the display. The Y coordinate is the distance from the top of the device, starting 25 millimeters above it and extending to about 600 millimeters (two feet) upwards. Note that the device cannot see below itself, so all Y coordinates will be positive.

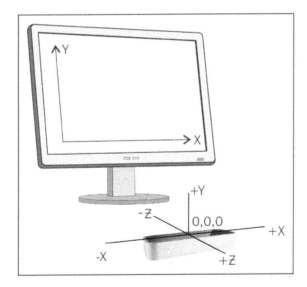

An example of cursor control

By now we are feeling pretty saucy, having diligently run the sample code thus far and controlling our computer in a way never before possible. While there is certain utility and endless amusement afforded by printing out finger coordinates while waving your hands in the air and pretending to be a magician, there are even more exciting applications waiting to be written, so let's continue onwards and upwards.

 Until computer-gesture interaction is commonplace, pretending to be a magician while you test the functionality of Leap SDK is not recommended in public places such as coffee shops.

In some cultures it is considered impolite to point at people. Fortunately your computer doesn't have feelings and won't mind if we use a pointing gesture to move its cursor around (you can even use a customarily offensive finger if you so choose). In order to determine where to move the cursor, we must first locate the position on the display that the user is pointing at. To accomplish this we will make use of the screen calibration and detection API in the SDK.

If you happen to leave your controller near a computer monitor it will do its best to try and determine the location and dimensions of the monitor by looking for a large, flat surface in its field of view. In addition you can use the complementary Leap calibration functionality to improve its accuracy if you are willing to take a couple of minutes to point at various dots on your screen. Note that once you have calibrated your screen, you should ensure that the relative positions of the Leap and the screen do not change.

Once your controller has oriented itself within your surroundings, hands and display, you can ask your trusty `controller` instance for a list of detected screens:

```
// get list of detected screens
const Leap::ScreenList screens = controller.calibratedScreens();

// make sure we have a detected screen
if (screens.empty()) return;
const Leap::Screen screen = screens[0];
```

We now have a screen instance that we can use to find out the physical location in space of the screen as well as its boundaries and resolution. Who cares about all that though, when we can use the SDK to compute where we're pointing to with the `intersect()` method?

```
// find the first finger or tool
const Leap::Frame frame = controller.frame();
```

```
const Leap::HandList hands = frame.hands();
if (hands.empty()) return;
const Leap::PointableList pointables = hands[0].pointables();
if (pointables.empty()) return;
const Leap::Pointable firstPointable = pointables[0];

// get x, y coordinates on the first screen
const Leap::Vector intersection = screen.intersect(
    firstPointable,
    true,   // normalize
    1.0f    // clampRatio
);
```

The vector `intersection` contains what we want to know here; the pixel pointed at by our pointable. If the pointable argument to `intersect()` is not actually pointing at the screen then the return value will be (NaN, NaN, NaN). **NaN** stands for **not a number**. We can easily check for the presence of non-finite values in a vector with the `isValid()` method:

```
if (! intersection.isValid()) return;
// print intersection coordinates
std::cout << "You are pointing at (" <<
    intersection.x << ", " <<
    intersection.y << ", " <<
    intersection.z << ")" << std::endl;
```

Prepare to be astounded when you point at the middle of your screen and the transfixing message **You are pointing at (0.519522, 0.483496, 0)** is revealed. Assuming your screen resolution is larger than one pixel on either side, this output may be somewhat unexpected, so let's talk about what `screen.intersect(const Pointable &pointable, bool normalize, float clampRatio=1.0f)` is returning.

The `intersect()` method draws an imaginary ray from the tip of `pointable` extending in the same direction as your finger or tool and returns a three-element vector containing the coordinates of the point of intersection between the ray and the screen. If the second parameter normalize is set to false then `intersect()` will return the location in the leap coordinate system. Since we have no interest in the real world we have set normalize to true, which causes the coordinates of the returned intersection vector to be fractions of the screen width and height.

When `intersect()` returns normalized coordinates, `(0, 0, 0)` is considered the bottom-left pixel, and `(1, 1, 0)` is the top-right pixel.

It is worth noting that many computer graphics coordinate systems define the top-left pixel as `(0, 0)` so use caution when using these coordinates with other libraries.

There is one last (optional) parameter to the `intersect()` method, `clampRatio`, which is used to expand or contract the boundaries of the area at which the user can point, should you want to allow pointing beyond the edges of the screen.

Now that we have our normalized screen position, we can easily work out the pixel coordinate in the direction of the user's rude gesticulations:

```cpp
unsigned int x = screen.widthPixels() * intersection.x;
// flip y coordinate to standard top-left origin
unsigned int y = screen.heightPixels() * (1.0f - intersection.y);

std::cout << "You are offending the pixel at (" <<
    x << ", " << y << std::endl;
```

Since `intersection.x` and `intersection.y` are fractions of the screen dimensions, simply multiply by the boundary sizes to get our intersection coordinates on the screen. We'll go ahead and leave out the Z-coordinate since it's usually (OK, always) zero.

Now for the *coup de grace* — moving the cursor location, here's how to do it on Mac OS X:

```cpp
CGPoint destPoint = CGPointMake(x, y);
CGDisplayMoveCursorToPoint(kCGDirectMainDisplay, de.stPoint);
```

You will need to #include `<CoreGraphics/CoreGraphics.h>` and link it (`-framework CoreGraphics`) to make use of `CGDisplayMoveCursorToPoint()`.

Now all of our hard efforts are rewarded, and we can while away the rest of our days making the cursor zip around with nothing more than a twitch of the finger. At least until our arm gets tired. After a few seconds (or minutes, for the easily-amused) it may become apparent that the utility of such an application is severely limited, as we can't actually click on anything.

So maybe you shouldn't throw your mouse away just yet, but read on if you are ready to escape from the shackles of such an antiquated input device.

A gesture-triggered action

Let's go all the way here and implement our first proper gesture—a mouse click. The first question to ask is, what sort of gesture should trigger a click? One's initial response might be a twitch of your pointing finger, perhaps by making a dipping or curling motion. This feels natural and similar enough to using a mouse or trackpad, but there is a major flaw—in the movement of the fingertip to execute the gesture we would end up moving the cursor, resulting in the click taking place somewhere different from where we intended. A different solution is needed.

If we take full advantage of our limbs, and assuming we are not an amputee, we can utilize not just one but both hands, using one as a "pointer" hand and one as a "clicker" hand. We'll retain the outstretched finger as the cursor movement gesture for the pointer hand, and define a "click" gesture to be the touching of two fingers together on the clicker hand.

Let's create a true Leap mouse application to support our newly defined clicking gesture. An important first step would be to choose a distance that represents two fingers touching. While at first blush a value of 0 mm would seem to be a reasonable definition of touching together, consider the fact that the controller is not always perfect in recognizing two touching fingers as being distinct from each other, or even existing at all. If we choose a suitably small distance we can call "touching" then the gesture will be triggered in one of the frames generated as the user closes their fingers together.

We'll begin with the obligatory listener class to handle frame events and keep track of our input state.

```
class MouseListener : public Leap::Listener {
public:
    MouseListener();
    const float clickActivationDistance = 40;
    virtual void onFrame(const Leap::Controller &);
    virtual void postMouseDown(unsigned x, unsigned y);
    virtual void postMouseUp(unsigned x, unsigned y);
    ...
};
```

On my revision 3 device, a distance of 40mm seems to work reasonably well.

For our `onFrame` handler we can build on our previous code. However now we need to keep track of not just one hand but two, which introduces quite a bit of extra complexity.

For starters, the method `Leap::Frame::hands()` is defined as returning the hands detected in a frame in an arbitrary order, meaning we cannot always expect the same hand to correspond to the same index in the returned `HandList`. This makes sense, because some frames will likely fail to recognize both hands and a new list of hands will need to be constructed as the detected hands are unrecognized and recognized again, and there is no guarantee that the ordering will be the same.

A further problem is that we will need to work out which is the user's left and right hands, because we should probably use the most dexterous hand as the pointer hand and the inferior, the clicker.

Indeed, even determining the primary and secondary hands is not quite as simple as one might think, because the primary and secondary hands will be reversed for left-handed people. Left-handed people have had it hard enough for thousands of years, so it would not be right for us to make assumptions.

> The English word "dexterity" comes from the Latin root *dexter*, relating to the right or right hand, and also meaning "skillful", "fortunate", or "proper" and often having a positive connotation. Contrast this with to the word for left—"sinister".

We'll start by adding some instance variables and initializers:

```
protected:
    bool clickActive; // currently clicking?
    bool leftHanded;  // user setting
    int32_t clickerHandID, pointerHandID; // last recognized
```

`leftHanded` will act as a flag which we can use when we determine which hand is the pointer and which is the clicker. `clickerHandID` and `pointerHandID` will be used to keep track of which detected hand from a given frame corresponds to the pointer and clicker.

We can create an initializing constructor like so:

```
MouseListener::MouseListener()
  : clickActive(false), leftHanded(false),
    clickerHandID(0), pointerHandID(0) {}
```

Explicitly initializing variables is good practice, in particular because the rules for which types are initialized in various situations in C++ are so multitudinous that memorizing them is discouraged. Using the initializer list syntax is considered good style because it can save unnecessary constructor calls when member objects are assigned new values, although since we are only initializing primitive types, we get no such reduction in overhead here.

```
Leap::Hand pointerHand, clickerHand;

if (pointerHandID) {
    pointerHand = frame.hand(pointerHandID);
    if (! pointerHand.isValid())
        pointerHand = hands[0];
}
```

If hands are detected, we will always at least have a pointer hand defined. If we've already decided on which hand to use (`pointerHandID` is set) then we should see if that hand is available in the current frame. When `Leap::Frame::hand(int32_t id)` is called with a previously detected hand's identifier, it will return a corresponding hand instance. If the controller has lost track of the hand it was following, then you'll still get a `Hand` back, but `isValid()` will be false. If we fail to locate our old hand or one hasn't been set yet, we'll assign the first detected hand for the case where we only have one hand in the frame.

```
if (clickerHandID)
        clickerHand = frame.hand(clickerHandID);
```

We attempt to locate the previously detected clicker hand if possible.

```
    if (! clickerHand.isValid() && hands.count() == 2) {
        // figure out clicker and pointer hand

        // which hand is on the left and which is on the right?
        Leap::Hand leftHand, rightHand;
        if (hands[0].palmPosition()[0] <= hands[1].palmPosition()[0])
{
            leftHand = hands[0];
            rightHand = hands[1];
        } else {
            leftHand = hands[1];
            rightHand = hands[0];
        }
```

Before we try to work out which hand is the clicker and which should be the pointer, we'll need to know which is the left hand and which is the right hand. A simple comparison of the X coordinates will do the trick nicely for setting the `leftHanded` flag.

```
if (leftHanded) {
    pointerHand = leftHand;
    clickerHand = rightHand;
} else {
    pointerHand = rightHand;
    clickerHand = leftHand;
}
```

Here we assign the primary hand to be the pointer, and the secondary to be the clicker.

```
clickerHandID = clickerHand.id();
pointerHandID = pointerHand.id();
```

Now that we've decided the hands, we need to retain references to those particular hands for as long as the controller can keep track.

```
const Leap::PointableList pointables = pointerHand.pointables();
```

Instead of `hands[0].pointables()` as before, now we'll want to use `pointerHand` for the screen intersection. The rest of the pointer manipulation code remains the same.

```
if (! clickerHand.isValid()) return;
```

Now it is time to handle the detection of a click, but only if there are two hands.

```
const Leap::PointableList clickerFingers =
clickerHand.pointables();
    if (clickerFingers.count() != 2) return;
```

If we don't find exactly two fingers on the clicker hand, then there is not going to be much we can do in terms of determining how far apart they are. We want to know if the user has touched two fingers together or not.

```
float clickFingerDistance = clickerFingers[0].tipPosition().
distanceTo(
clickerFingers[1].tipPosition()
);
```

The `Leap::Vector` class has a handy `distanceTo()` method that tells us how far apart two points in space are.

```
    if (! clickActive && clickFingerDistance <
clickActivationDistance) {
        clickActive = true;
        cout << "mouseDown\n";
        postMouseDown(x, y);
```

If we have not already posted a mouse down event and if the clicker hand's two fingers are touching, then we will simulate a click with `postMouseDown()`.

```
    } else if (clickActive && clickFingerDistance >
clickActivationDistance) {
        cout << "mouseUp\n";
        clickActive = false;
        postMouseUp(x, y);
    }
```

And likewise for when the two fingers come apart, we finish the click and release the button. Unfortunately, just as with the cursor movement code, there is no simple cross-platform way to synthesize mouse events, but the OSX code is provided as follows for completeness:

```
void MouseListener::postMouseDown(unsigned x, unsigned y) {
    CGEventRef mouseDownEvent = CGEventCreateMouseEvent(
                            NULL, kCGEventLeftMouseDown,
                            CGPointMake(x, y),
                            kCGMouseButtonLeft
                                      );
    CGEventPost(kCGHIDEventTap, mouseDownEvent);
    CFRelease(mouseDownEvent);
}

void MouseListener::postMouseUp(unsigned x, unsigned y) {
    CGEventRef mouseUpEvent = CGEventCreateMouseEvent(
                            NULL, kCGEventLeftMouseUp,
                            CGPointMake(x, y),
                            kCGMouseButtonLeft
                                      );
    CGEventPost(kCGHIDEventTap, mouseUpEvent);
    CFRelease(mouseUpEvent);
}
```

And now you can throw away your mouse for good! Actually, don't do that. First be sure to run the screen calibration tool.

Truth be told, there are plenty of improvements that could be made to our simple, modest mouse replacement application. Implementing right-click, a scroll wheel and click-and-drag are left as an exercise for the reader.

Summary

And thus begins our exciting "leap" into the SDK, starting with the basics of reading finger information and tracking where the user is pointing. We'll continue filling in the details of the rest of the functionality in the SDK along with some more fun examples in the rest of the book. While we have only just scratched the surface of what can be done with Leap, you should already be starting to get ideas of how to engage with users using hand motion input. Try out some of the example applications that come with the SDK for inspiration to get a feel for what it can do.

You should now have a working application written in C++ with a frame callback that has access to all of the hand tracking data captured by the controller. Next up, we'll look at making an application interface that is as responsive as possible.

2
Real Talk – Real Time

When we create an application that acts on gestural input, a major concern is responsiveness. Not only is using a laggy application no fun, but there are many situations in which we desire to process and respond to the hand motion with as little latency as possible. In this chapter we'll discuss the following:

- Blocking
- Describing raw input as simple gestures
- Shared-memory concurrency with pthreads
- Building a MIDI controller for real-time audio applications

With this in mind, we will walk through creating a controller suitable for live musical performances. Because our code will be interacting with music software and hardware, we must use our operating system interfaces for transmitting MIDI (the standard musical control protocol that all devices support), which is not guaranteed to happen without blocking. While using routines that can potentially involve sending messages to hardware, it's generally good to assume that the execution of our program may be paused while reception or transmission is taking place. We will need an efficient way to handle gestures triggered by the user while sending our output through a blocking interface at the same time.

 Blocking is when execution of a thread of code is stopped for a period of time while waiting on a lengthy action to complete. Many networking and file access routines block by default, which makes writing linear programs simpler at the expense of preventing your program from doing more than one thing at a time.

Consider the state of modern electronic music performances and DJing; these days you're just as not likely to see anything more going on than an unkempt dude in a hoodie who simply presses a couple of buttons on his keyboard. Not very exciting to watch. Why shouldn't they have the ability to use their controller to dazzle the audience with intricate and wild hand motions, as if they are the conductor of a one-laptop orchestra? Let's see if we can spice things up a little, with our own Leap-based MIDI controller.

If you've ever done any work with electronic musical instruments such as a keyboard or sequencing software, you have likely run into **Musical Instrument Digital Interface (MIDI)**. If you want to plug your keyboard into a synthesizer program, or drum pads into a drum machine, or a bunch of knobs and sliders to a live performance application, or a dishwasher into a robot-controlled xylophone, MIDI is the way to go. The protocol makes it trivial to emit and receive standard messages such as notes (including pitch, volume, and more) and control messages (a numeric value between 0 and 127) that can be mapped to anything your heart desires. If you wish to have a note trigger a choir or the sound of one hand clapping, there is plenty of software and hardware that makes it a snap.

To cut a long story short, if we make a program that recognizes some simple gestures and outputs MIDI messages corresponding to hand movements, we can enable artists to give a fascinating performance worth watching.

Before we sail off on our voyage across the implementation ocean, we must first map out our journey with the interface harbormaster. Let us ponder a couple of questions:

- What will our program do?
- How will it do it?

The first question should be straightforward enough to answer; we want to allow our users to use the controller to interact with their music software and hardware. The most straightforward way to accomplish that is by sending MIDI control messages, which can be mapped by the user to any parameter they desire, such as a filter cut-off frequency or the pitch of a synthesizer. One might think of it as an improvement on the classic Theremin instrument, which was an analog instrument invented in the 1920s that was controlled by hand motions. The limitation of the Theremin design is that it outputs audio, and therefore, can only produce one type of sound, whereas our MIDI controller can be made to control any type of instrument or effect parameter.

In our program, there should be a set of predefined motions that trigger distinct MIDI control messages, which are defined as a controller number (between 0 and 127) and a value to set for that controller. Thus, we could define the motion of a fingertip as three distinct controller numbers, one for each axis of motion, and the values corresponding to the coordinates in space.

Now that we have a solid idea of the purpose of our program, how do we design a C++ program to accomplish this? Let's begin by defining some types that we can work with.

```cpp
namespace leapmidi {

typedef double midi_control_value_raw;
typedef unsigned short midi_control_value;
```

One of the main functions of our program is going to be mapping between raw input data from the controller in the form of coordinates (given in millimeters, represented as double) to a MIDI control value, which is a number between 0 and 127.

```cpp
// MIDI controller numbers
enum midi_control_index {
    BALL_RADIUS_HAND_1 = 1,
    BALL_RADIUS_HAND_2,

    FINGER_1_HAND_1_X,
    FINGER_1_HAND_1_Y,
    FINGER_1_HAND_1_Z,
    FINGER_2_HAND_1_X,
    FINGER_2_HAND_1_Y,
    FINGER_2_HAND_1_Z,

    FINGER_1_HAND_2_X,
    FINGER_1_HAND_2_Y,
    FINGER_1_HAND_2_Z,

    // ...
};
```

Here we statically define a controller number for a few basic sorts of events.

```cpp
class Gesture;
class Control;
typedef std::shared_ptr<Control> ControlPtr;
typedef std::shared_ptr<Gesture> GesturePtr;
} // namespace leapmidi
```

And we'll polish off our main header file with some forward declarations of the Leap MIDI `Gesture` and `Control` classes, and typedef shared pointers for easier management of memory and reference passing.

It won't do to simply forward declare `leapmidi::Gesture` and `leapmidi::Control` and leave the compiler hanging, so let's sketch those out next. What's in a `Control` class anyway? Well, we should keep track of what caused the control and what the raw input from the controller was, so that we can output the appropriate control number and value.

```
namespace leapmidi {

class Control {
public:
    Control(midi_control_value_raw rawValue, int hand) :
        _rawValue(rawValue), _handIndex(hand) {}
```

Here we'll keep track of two instance variables, the one that hand-triggered the control and what the value of the motion was (for example, position, velocity, or rotation). A way to convert from a raw input value to a normalized output in the range of 0-127 is required, and to perform such an operation, the range of possible input values will need to be defined by child classes.

```
    // min/max raw value, for mapping to MIDI value
    virtual midi_control_value_raw minRawValue() = 0;
    virtual midi_control_value_raw maxRawValue() = 0;
    // map a raw value from [minRawValue,maxRawValue] into the range
[0,127]
    virtual midi_control_value mappedValue();
```

The last piece of the MIDI output puzzle is to get our controller number. As with `minRawValue()` and `maxRawValue()`, we can force subclasses to define a value by creating a pure virtual method.

```
    virtual midi_control_index controlIndex() = 0;
```

Defining a human-readable description of the control is always handy for debugging.

```
    virtual const std::string &description() = 0;
```

Adding the instance variables should round things out nicely.

```
    // public accessors
    int handIndex() { return _handIndex; }
    midi_control_value_raw rawValue() { return _rawValue; }
```

```
protected:
    int _handIndex;
    midi_control_value_raw _rawValue;
}; // class Control


} // namespace leapmidi
```

Reflecting on our shiny new `Control` class, it seems that we've defined the majority of the internal interface for our program; we can stick in a value obtained from a user interaction and compute our MIDI output value and controller number.

In addition to structuring our program's outputs, it is a good idea to define what our gesture inputs should look like. If we create a base class for interpreting hand motions, we can easily partition our code up into small modules for each type of gesture, and we can also pick and choose which gestures we want to be active at a given time.

Simple enough, right? Yes it is.

```
class Gesture {
public:
    // given a controller, return MIDI controls recognized from
gestures
    virtual void recognizedControls(const Leap::Controller
&controller, std::vector<ControlPtr> &gesturesOut) = 0;
};
```

The `Gesture` class encapsulates the functionality to transform a `Leap::Frame` into a series of controls. Two things are required to make the magic happen here: a controller is needed so that our recognizer can examine the current and past frames to determine if a gesture that it recognizes has been triggered, and a list to store the control messages that this gesture has spat out. Filling in a vector of shared pointers supplied to your method is one way of returning a list of objects, although this is not the only possible solution here. It has the advantage of not requiring any extra copying, as opposed to returning a vector.

Why bother distinguishing between a gesture and a control in the first place? Because a gesture, such as a fingertip movement, can result in multiple control messages. In the case of our fingertip movement gesture, we might as well emit separate messages for fingertip's X position, Y position, and Z position, so that we can map each axis to a different parameter.

To make things actually work, we'll need something to create some gesture recognizers and update them when a new input is received. This sounds like a job for our trusty sidekick, the listener subclass! To make our program a little bit more modular and event-driven, we will define some additional callbacks that will be invoked when input of interest to us takes place.

```
namespace leapmidi {

class Listener : public Leap::Listener {
public:
    Listener();

    // called when we have identified a gesture in the current frame
    virtual void onGestureRecognized(const Leap::Controller
&controller, GesturePtr gesture);
```

When we detect a predefined gesture, say a fingertip movement, we'll pass the detected gesture to this callback. This won't be immediately useful for our purposes, but is included for completeness, and this would certainly be a good location for adding the debug output and user feedback. One of the recommendations from the SDK user experience guidelines is to concentrate on providing dynamic feedback of actions to your users, so that they can interact with your application with more precision.

```
    // called for each control message emitted by a gesture recognizer
    virtual void onControlUpdated(const Leap::Controller &controller,
GesturePtr gesture, ControlPtr control);
    // vector of Gesture instances to detect gesture input
    // and emit control messages
       std::vector<GesturePtr>& gestureRecognizers() { return _
gestureRecognizers; }
```

onControlUpdated() will be our callback for control messages that should be sent to the MIDI output.

Our listener will need to keep track of the gestures it is currently interpreting. Also, by making a dynamic list of recognizers, a future version of our program could define a number of different sets of recognizers and outputs that the user could toggle between to allow for different modes of operation.

```
    // from Leap::Listener
    virtual void onFrame(const Leap::Controller &controller);

protected:
    std::vector<GesturePtr> _gestureRecognizers;
```

Here we will revisit our old friend, `onFrame()`, in which we will analyze the current frame for any gestures that we can recognize. With the declaration of our current recognizers, we can safely set sail across the great implementation ocean.

```
void Listener::onFrame(const Leap::Controller &controller) {
    // feed frames to recognizers
    vector<GesturePtr> recognizers = gestureRecognizers();
    for (vector<GesturePtr>::iterator it = recognizers.begin(); it !=
recognizers.end(); ++it) {
```

When the Leap driver has new input data, we will want to inform all our recognizers via the `recognizedControls()` method on each recognizer.

```
    for (vector<GesturePtr>::iterator gesture = recognizers.begin();
gesture != recognizers.end(); ++gesture) {
        vector<ControlPtr> gestureControls; // controls from this
gesture
        (*gesture)->recognizedControls(controller, gestureControls);

        if (! gestureControls.size())
            continue;
```

Asking each recognizer to cough up some control outputs is as straightforward as iterating over our list of recognizers and calling `recognizedControls()` on each, providing an empty list in which it can stash the resulting controls.

```
        onGestureRecognized(controller, *gesture);
```

Since we didn't hit the continue statement in the preceding line of code, we can infer that some useful information has been extracted from the current frame. Now would seem to be a good time to call the `onGestureRecognized()` callback that we made up in our `Listener` class.

```
        for (vector<ControlPtr>::iterator ctl = gestureControls.
begin(); ctl != gestureControls.end(); ++ctl) {
            onControlUpdated(controller, *gesture, *ctl);
        }
    } // for gestures
} // onFrame
```

Now that we've got our gesture and our controls, let's do something fun and awesome with them.

```
// do something productive with these in your application's Listener
subclass
void Listener::onGestureRecognized(const Leap::Controller &controller,
GesturePtr gesture) {}
```

```
void Listener::onControlUpdated(const Leap::Controller &controller,
GesturePtr gesture, ControlPtr control) {
    cout << "recognized control index " << control->controlIndex()
    << " (" << control->description() << ")"
    << ", raw value: "
    << control->rawValue() << " mapped value: " << control-
>mappedValue() << endl;
}
```

Okay so I lied. We'll come back to this soon and deal with our MIDI output.

Now that we've got this code to deal with gesture recognizers in their generic, shapeless form, it's about time we created a real recognizer.

A simple gesture recognizer

One of the handy methods we can call on a `Leap::Hand` is `sphereRadius()`, which returns the size of an imaginary ball filling up a hand. The more outstretched your fingers are the larger the radius, and a more clenched fist will produce smaller values.

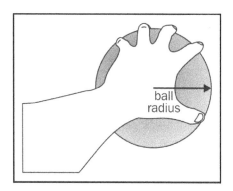

What does our ball gesture recognizer class look like? Feast your eyes on this:

```
typedef std::shared_ptr<BallGesture> BallGesturePtr;

void BallGesture::recognizedControls(const Leap::Controller
&controller, std::vector<ControlPtr> &controls) {
    Leap::Frame frame = controller.frame();
    if (frame.hands().empty())
        return;
```

If no hands are found in the current frame, there's not much else for us to do here. However, if the user is waving some appendages around, we can do our thing.

```
for (int i = 0; i < frame.hands().count(); i++) {
    if (i > 1) break;
```

If one or two hands are detected in the frame, we might be able to do something useful. We can get extra fancy if we return a different control message depending upon how many hands are held out, doubling the usefulness of this gesture.

If more than two hands are detected, it may mean that more than one person (or one mutant) is operating the controller, but most likely it means that the Leap driver software is not totally perfect.

```
Leap::Hand hand = frame.hands()[i];
double radius = hand.sphereRadius();
if (! radius)
    continue;
```

Here we get the radius in millimeters of the imaginary ball held by this hand. Much of the time that the controller will be unable to come up with a satisfactory answer and will just give you a zero instead, so let's ignore that.

```
BallRadiusPtr bc = make_shared<BallRadius>(radius, i);
ControlPtr cptr = dynamic_pointer_cast<Control>(bc);
```

Now that we've figured out how many hands we are dealing with and their radii, we can go ahead and construct an instance of a `BallRadius` control, which will be defined momentarily. If you recall, the `Control` constructor takes the raw input value and the hand index as arguments.

Time to return our freshly minted `Control` by sticking it in vector of the `Control` classes, right? Not so fast, the compiler may be aware that `BallRadius` is a subclass of `Control`, but unfortunately such class hierarchy of information is lost while using shared pointers. Using `dynamic_pointer_cast` will prevent a syntactical nastygram by assuring the compiler that we can safely upcast `BallRadius` to `Control` when we add it to the list of controls to return from inside `BallGesture::recognizedControls`:

```
controls.push_back(cptr);
```

Good stuff. Now we need a ball radius control.

```
class BallRadius : public Control {
public:
```

```
    BallRadius(midi_control_value_raw rawValue, int hand) :
Control(rawValue, hand) {};

    virtual const std::string& description() {
        static const std::string desc = "Hand curvature";
        return desc;
    }


    virtual midi_control_value_raw minRawValue() { return 48; }
    virtual midi_control_value_raw maxRawValue() { return 115; }
```

Roughly speaking, 48-115 is the range of raw values returned by `sphereRadius()` as determined by experimentation. One day the Leap SDK may provide a way of obtaining a normalized value, but this will serve our purposes for now.

```
    virtual midi_control_index controlIndex() {
        if (handIndex() == 0) return BALL_RADIUS_HAND_1;
        if (handIndex() == 1) return BALL_RADIUS_HAND_2;
        throw "Too many hands detected for ball radius control!";
    }
}
```

If we want each hand to emit its own control message, we can simply assign them different control numbers. These come from the `midi_control_index` enum.

```
    typedef std::shared_ptr<BallRadius> BallRadiusPtr;
```

To make things a little bit cleaner in the previously described gesture recognizer, we'll make a type for referring to a `shared_ptr` for our control.

We are very nearly ready to test our program out now. One final detail to fill in is the mapping from a `Control` class's raw input value to its MIDI control message output value, which you recall must be in the range of 0-127.

```
bool _controlRangeWarning = true;

midi_control_value Control::mappedValue() {
    midi_control_value_raw min = minRawValue();
    midi_control_value_raw max = maxRawValue();
    midi_control_value_raw raw = rawValue();

    if (raw < min) {
        if (_controlRangeWarning)
            cerr << "Warning, " << description() << " got raw value "
<< raw
```

```
                << " which is below the min of " << min << std::endl;
        raw = min;
      } else if (raw > max) {
          if (_controlRangeWarning)
              cerr << "Warning, " << description() << " got raw value "
  << raw
                << " which is above the max of " << max << std::endl;
          raw = max;
      }
```

Since we are deriving the possible range of input values from experimentation, it would be good to know if we receive a new value that is outside our currently defined minimum and maximum range.

```
    midi_control_value_raw inValNorm = raw - min;
    midi_control_value_raw upperNorm = max - min;
    midi_control_value_raw normPosition = inValNorm / upperNorm;
    midi_control_value_raw outVal = normPosition * 127;
    return outVal;
}
```

This formula will convert values in the range of (min, max) to the range of (0, 127), giving us the final piece of data needed to generate our MIDI output.

To recap: our program has a Listener to capture motion events from the controller, gesture recognizers to feed the motion events, callbacks triggered by recognized gestures, and control messages being emitted with all the data required to interface with other applications and hardware. Not only that but we have a fairly clean, modular application that we can add features to in a noninvasive fashion. Not bad for just a few pages of code!

It is time to reward our tireless efforts with a test run. To demo the software place your hands over the controller and pretend as though you are picking up some oranges. You will no doubt be overcome with boundless joy, as you gaze upon your console messages and see the following output:

```
recognized control index 1 (Hand curvature), raw value: 109.715 mapped
value: 116
recognized control index 2 (Hand curvature), raw value: 82.577 mapped
value: 65
```

If we were an agile development team, we could tag our software with a 1.0 release version number and ship it at this point. Instead, we are going to hook it up as a virtual MIDI device. To do that we are going to need a grip of instance variables.

```
MIDIClientRef deviceClient;
MIDIEndpointRef deviceEndpoint;
MIDIPacketList *midiPacketList;
MIDIPacket *curPacket;
```

And we should probably initialize them or something I guess.

```
MIDIClientCreate(CFSTR("LeapMIDI"), NULL, NULL, &deviceClient);
MIDISourceCreate(deviceClient, CFSTR("LeapMIDI Control"),
&deviceEndpoint);
```

Using the MIDI output

A MIDI client is an entity that can communicate with other MIDI clients, and may have one or more inputs and outputs. Our client will be sort of useless without any I/O, so we create a virtual source into which we can inject our freshly minted MIDI messages. The descriptive string passed in will appear in the MIDI configuration dialogs of any application that supports MIDI, such as my DAW of choice, Ableton Live.

Let's get to the meat of sending a MIDI control message now; we will deal with the packet initialization in a minute. Remember our onControlUpdated() routine? We can use the mapped control value and the controller index number to build a message to transmit.

```
midi_control_index ctrlIdx = control->controlIndex();
midi_control_value ctrlVal = control->mappedValue();
```

ctrlIdx and ctrlVal correspond to the number of the MIDI control and the output value for that control respectively. For our ball radius gesture, the control indices will be BALL_RADIUS_HAND_1 and BALL_RADIUS_HAND_2, defined by our enum as 1 and 2.

```
assert(ctrlIdx <= 119);
assert(ctrlVal <= 127);
```

MIDI control change messages are limited up to 119; higher values have special predefined meanings such as turning off all notes, resetting controllers, and changing between mono- and polyphonic-note mode (this determines if more than one note may be played at once).

```
if (midiPacketList)
    free(midiPacketList);
```

```
    midiPacketList = (MIDIPacketList *)malloc(packetListSize *
sizeof(u_int8_t));
    curPacket = MIDIPacketListInit(midiPacketList);
```

The OS X MIDI API requires us to reinitialize an array of packets to transmit for each collection of messages. Note that `malloc()` and its memory management pals have potentially unbounded execution time and can potentially block execution (though this is more of a theoretical concern).

Once our packet list is reinitialized, we can create a new MIDI message to add to it.

```
// build midi packet
u_int8_t packetOut[3];
packetOut[0] = 0xB0;     // channel
packetOut[1] = ctrlIdx;  // controller number
packetOut[2] = ctrlVal;  // controller value
```

`0xB0-0xBF` is defined by the MIDI specification as controller change messages on different channels, with `0xB0` being channel 1, `0xB1` being channel 2, and so on. Supporting input and output on multiple channels allows an application or device to support multiple distinct connections or configurations, which is more complex than we need. `0xB0` should suit us just fine for now.

Control change messages consist of three bytes: the control change opcode and channel offset, the controller number of function identifier, and the value to change the controller to. `packetOut[1]` is the number of our control, as given by the ball radius control ID, and `packetOut[2]` is our mapped output value.

Now we have our message packet and we can append it to the list of messages to transmit.

```
MIDITimeStamp timeStamp = mach_absolute_time();
curPacket = MIDIPacketListAdd(midiPacketList, packetListSize,
curPacket, timeStamp, 3, packetOut);
```

That's quite a mouthful, what are all those variables again? `curPacket` is a pointer indicating where the last value in the list was, so that when we append more values, they will go after the previous packets instead of overwriting whatever is at the start of the packet list.

`packetListSize` is a constant that can be anything really; as long as we are not going to fill it up with packets waiting to be sent, `1024` is a perfectly fine value. If we ever try to append packets with a full list and an insufficient `packetListSize` value, the return value here will be `NULL` and we will know that our `packetListSize` value is too small (because we always check our return values, even if the checks are often elided in this guide for the sake of brevity).

`timeStamp` is there for keeping track of when a MIDI message was received. As we are creating a virtual MIDI device source and pretending that the controller is an instrument, technically we are receiving this message from the perspective of the MIDI system.

Lastly, the lame hardcoded 3 corresponds to the number of bytes in `packetOut`, the packet we are appending to the packet list. This is the same as the line `u_int8_t packetOut[3]` seen in one of the earlier code snippets.

Where are we at now? There's an MIDI device ready and waiting and a list of packets to feed into its gaping maw. How do we send them? This can be done by calling the `MIDIReceived()` function. (The fact that Apple does not really document Core MIDI makes figuring out these function names a really exciting game.)

```
MIDIReceived(deviceEndpoint, midiPacketList);
```

Remember that we are not sending our messages from the computer; we must indulge the MIDI system and pretend that the virtual Leap MIDI device is receiving control messages as if it were plugged into an instrument.

Blocking and latency

So, can we play the Leap now? We absolutely can! There is a wealth of free and trial music software available for generating all sorts of noises a quick Google search away. You should be able to select the Leap MIDI control device in your music program's settings, and use its **MIDI learn** feature to map your ball radius gestures to anything you desire (filter cut-offs, effect parameters, and synthesizer knobs are especially fun).

This is all well and good but there is one very serious, unaddressed issue with our MIDI controller: **blocking**. The sad truth is that our code is not ready to be used by the general public to give performances because the gesture response time is utterly nondeterministic. In the `onControlUpdated()` routine described earlier, our call to `MIDIReceived()` is part of a poorly documented API that may or may not involve communication with blocking subsystems or hardware, and that frankly we know very little about. Depending upon the user's setup, hardware, system load, and versions of involved libraries, the sending of MIDI control data may significantly delay the processing of new frames of input data and the responsiveness of our program, which is a literally show-stopping calamity. Not only will it prevent any other processing from taking place, but it will also cause any interface our program has to become unresponsive and delay the recognition of new gestures.

Our MIDI controller was not just a demonstration of a nifty practical use of the controller, but also an example of a major latency issue that anyone developing a Leap-enabled program is likely to face. You will probably enter a world of pain if your app performs synchronous file or network I/O, having an interactive user interface, is running on a CPU with multiple cores, or needs to call functions that are unlikely to have execution time ceilings (such as our good friend `MIDIReceived()`, and even possibly `malloc()` and `free()`).

The main thrust of the problem is that we need to be continuously receiving our `onFrame()` callbacks while our application is busy performing other tasks that could potentially interfere with the regular and rapid processing of frames. Without proper multiprocessing of some sort, we run the risk of acting on gestures late, which in our case of a live audio controller would be even worse than dropping the input entirely.

Because we are continuously sending MIDI update messages, losing a few frames of data is acceptable because we will just send a new update with nearly identical data almost immediately. In cases such as ours, it is more important to make sure our program's output reflects the current state of the users' desires than to always handle every frame. Imagine a DJ trying to crossfade songs timed on the beat, but the controller is still trying to work through frames from many milliseconds ago to catch up instead of simply responding to the latest hand position.

Multiprocessing with threads

The solution to our conundrum is to run more than one thread of execution at the same time. CPUs and modern operating systems are very good at dealing with multiple processes and multiple threads of execution within processes, and this suits our purposes nicely. What we really want is one task that is receiving callbacks from the controller and handling the input, with a separate task taking our control messages and outputting them. This would give us a robust setup in which we can easily do anything we want in the output thread, without introducing issues associated with blocking the input processing thread. We could be happily logging output to a file, sending it over a network, outputting MIDI to hardware, controlling GPIO pins, doing expensive calculations or 3D graphics, and not affect the response time of our Leap callbacks.

As with many things in software development, there is more than one way to achieve our desired result of multitasking: POSIX threading, Boost threads, forking, asynchronous I/O, and so on. Forking would complicate our system considerably with process management code and limit shared access of program state, and the MIDI output subsystem has no documented capability to work asynchronously; so we shall turn our attention to threads.

POSIX threading (pthreads) is a widely used part of the **Portable Operating System Interface** standard that defines multithreading primitives such as mutexes, condition variables, and threading. Any Unix-like system such as Linux or Mac OS X is at least mostly POSIX-compliant, as is Windows via the **Windows Services For UNIX** component. While pthreads may lack some of the fancy C++ constructs wielded by the Boost threading library, it should prove to be quite sufficient for our needs without introducing too much complexity or a dependency on any third-party libraries. pthreads by its simplistic nature is well suited to a demonstration of the multithreading fundamentals that we require.

Refactoring for multithreading

Before continuing further on, perhaps it would be a good time to refactor our virtual MIDI device output functionality into its own class, as the MIDI output routines with the attendant threading code should be encapsulated for maintaining some semblance of clean program structure.

```
namespace leapmidi {
class Device {
public:
    Device();
    ~Device();
    void init();
    void queueControlMessage(midi_control_index controlIndex, midi_
control_value controlValue);
```

We'll go ahead and create a `leapmidi::Device` class. Users of our class construct a new `Device` class and call `init()` to set up the MIDI virtual device, and then send MIDI control messages using `queueControlMessage()`.

```
protected:
    void initPacketList();
    void createDevice();
```

Our earlier MIDI virtual device setup code can go in here.

```
    pthread_t messageQueueThread;
    void *messageSendingThreadEntry();
```

This function will be called when our MIDI output thread (kept track of by `messageQueueThread`) begins, and it will loop forever until the `Device` class is destroyed. Its job will be to sit and wait for new control messages to output, and then emit them using `MIDIReceived()`.

```
    std::queue<midi_message> midiMessageQueue;
    pthread_mutex_t messageQueueMutex;
    pthread_cond_t messageQueueCond;
```

Here we define a queue of `midi_messages` that are standing by, waiting to be sent out. Access to the queue will be guarded by the `messageQueueMutex` mutex and notifications of new queue items will be broadcast via the `messageQueueCond` condition variable. We need to keep track of messages in the queue, so we'll need a container to stick our control messages in. A simple `struct midi_message` type can easily contain the message info.

```
typedef struct {
    midi_control_index control_index;
    midi_control_value control_value;
    timeval timestamp;
} midi_message;
```

We will also need the assortment of the MIDI device and packet list members from before, along with some protected methods for adding MIDI packets and transmission.

```
    void _addControlPacket(midi_control_index control, midi_control_
value value);
    void _addControlMessages(std::queue<midi_message> &messages);

    // send midi packets
    OSStatus _flushMessages();

    MIDIClientRef deviceClient;
    MIDIEndpointRef deviceEndpoint;
    MIDIPacketList *midiPacketList;
    unsigned int packetListSize;
    MIDIPacket *curPacket;
```

Last and very much assuredly least, we will need a way to use an instance method as our thread entry point, the recently mentioned `messageSendingThreadEntry()` class to be precise. Here, we encounter one of the downsides of using a C++ class with a threading library designed for C applications, although it fortunately only entails a very small bit of the wrapper code, which we can write and then never think about again.

```
private:
    static void *_messageSendingThreadEntry(void *This) {
        ((Device *)This)->messageSendingThreadEntry(); return NULL;
    }
```

pthread entry points are passed in a `void *` parameter on their creation; so when we create our thread, we can simply pass in the current instance and then call our "real" entry point method (`messageSendingThreadEntry()`) on that. Let's do that in our `init()` method, called when our `Device` client is ready to begin.

```
void Device::init() {
    initPacketList();
    createDevice();

    // start message sending queue
    int res = pthread_create(&messageQueueThread, NULL, _
messageSendingThreadEntry, this);
    if (res) {
        cerr << "pthread_create failed " << res << endl;
        exit(1);
    }
}
```

Here we kick our thread off with `pthread_create()`, and pass `this` as our parameter to `_messageSendingThreadEntry()`.

Before we get to the thread entry point, we should have a look at how MIDI control messages are passed between threads. What we know is that the main thread in which our Leap callbacks such as `onFrame()` should do as little heavy lifting as possible, in order that it may continue receiving timely updates. Ideally, `onFrame()` should just feed the message sending thread MIDI messages to output, which we can accomplish by means of a queue of pending messages and the public `queueControlMessage()` method.

```
void Device::queueControlMessage(midi_control_index controlIndex,
midi_control_value controlValue) {
    midi_message msg;
    msg.control_index = controlIndex;
    msg.control_value = controlValue;
    gettimeofday(&msg.timestamp, NULL);

    pthread_mutex_lock(&messageQueueMutex);

    // critical section
    midiMessageQueue.push(msg);

    pthread_mutex_unlock(&messageQueueMutex);
    pthread_cond_signal(&messageQueueCond);
}
```

Here we introduce a critical section: we must make concurrent access to the `midiMessageQueue` mutually exclusive. To allow one thread to read from the queue while the other is in the process of modifying, it would result in all sorts of pain and suffering, so we need to protect the queue with a mutex while pushing our message onto it. Note that we keep the critical section as small as possible to limit the amount of time that is spent holding the mutex, to prevent the other thread from blocking if it is trying to acquire the lock.

After we have finished our business of adding the message to send to the queue, we signal on a condition variable, which is like a thread poking stick. If the MIDI message sending thread is not busy sending a message, it will be sitting around waiting for more messages to be added to the queue. In this case, we want the new message on the queue to be processed in as timely a manner as possible, so we poke the sleeping thread by signaling the condition variable. `pthread_cond_signal()` is used to wake up a thread that is listening on a condition variable.

Oh yeah, we should probably initialize the mutex and condition variables before using them.

```
Device::Device() {
    pthread_mutex_init(&messageQueueMutex, NULL);
    pthread_cond_init(&messageQueueCond, NULL);
    … more initialization
```

Fantastic. We are now ready to implement the meat of our message sending thread.

This thread will need to perform several tasks: it will need to check for new messages to transmit or wait for new messages, it will need to copy the messages from the queue to some thread-local storage, and then lastly it will need to emit the MIDI messages.

We can check to see if messages are waiting to be sent by seeing if `midiMessageQueue` is empty. As we just saw, this is our protected conduit between our two threads, so access to it should be protected with the `messageQueueMutex` lock.

```
void *Device::messageSendingThreadEntry() {
    while (1) {
        pthread_testcancel();
        pthread_mutex_lock(&messageQueueMutex);
        if (midiMessageQueue.empty()) {
```

Our thread will chill in an infinite loop as it waits for messages. We can cancel the message sending thread in the destructor when our Device instance is destroyed, so we check to see if the thread has been cancelled with `pthread_testcancel()`. If the thread was cancelled, `pthread_testcancel()` will never return and instead will cleanly terminate the thread, so it is a good idea to place it in a part of your code where it is safe to jump out.

After checking to see if our thread is still going on, we want to see if there are messages waiting in the queue. Before we can peek in `midiMessageQueue` we will need to acquire the lock, in order to be sure that our other thread is not in the middle of adding anything to the queue.

If we find messages in the queue, great; we will deal with them in just a minute. Supposing we don't though, then what? Here things get a bit tricky.

The producer-consumer race condition

What we have created is a classic producer-consumer pattern, where one thread is producing work and the other is processing the work on the queue. We are trying to make our application take care of the following:

- Respond to events in as close to real-time as possible
- Not lose any events

So, we need to be absolutely certain that we handle new items added to the queue as rapidly as possible. A naive implementation might look something like this:

```
- release message queue lock
- (possible race condition here!)
- wait for condition variable broadcast
- check queue again
```

But that would be utterly wrong. Suppose that `queueControlMessage()` is called *after* unlocking our message queue mutex, but *before* our thread starts waiting for the main Leap processing thread condition variable signal. The condition variable signals are ephemeral in the extreme. If no thread is waiting at the exact moment the signal is sent, we will never know that our thread has been poked. Because we just checked to see if the queue is empty but haven't gotten around to waiting for the signal yet, we very well could end up not handling that message until it is much later. Well, not that much later, but we care a lot about handling these events in a timely fashion, so we want to eliminate any possible delays.

This is a classic producer-consumer program, and it is a fundamental obstacle that requires a particular sort of atomic function to effectively wait for a condvar signal and lock a mutex at the same time. `pthread_cond_wait()` and `pthread_cond_timedwait()` are handy functions that atomically begin listening for condition variable signals at the same time that they unlock our mutex. Because these two operations effectively happen simultaneously, there is no danger of our execution being interrupted and the main thread poking our condition variable before we have started listening for the poke.

```
        int res = pthread_cond_wait(&messageQueueCond,
&messageQueueMutex);
        if (res != 0) {
            cerr << "unexpected pthread_cond_wait retval=" << res
<< endl;
            exit(1);
        }
```

Here, we release our lock on messageQueueMutex as soon as our thread starts waiting for a signal on messageQueueCond. When a signal is raised, pthread_cond_wait() reacquires the message queue mutex, and we can finally look in our queue to see if there is work to be done. At this point midiMessageQueue should have stuff in it for us to process.

Now is it safe to send the contents of our message queue out through our MIDI device? Sadly no, because as we just learned when pthread_cond_wait() returns, the mutex lock will be held by our thread again. If we do our thing and call MIDIReceived() now, all our hard work will be for naught, that is, we will end up blocking while holding the message queue mutex, which will block the main Leap callback thread. We want to do as little work as possible in this critical section where we have the mutex locked. So, let's quickly copy our messages out of the queue and into our local thread copy, empty it out, and then unlock and deal with the messages afterwards.

```
queue<midi_message> queueCopy;
while (! midiMessageQueue.empty()) {
    midi_message msg = midiMessageQueue.front();
    queueCopy.push(msg);
    midiMessageQueue.pop();
}
// unlock
if (pthread_mutex_unlock(&messageQueueMutex) != 0) {
    cerr << "message queue mutex unlock failure\n";
    exit(1);
}
```

Copying the messages to storage local to the thread isn't a particularly expensive operation, especially since there should only be a trivial number of MIDI messages waiting to be sent. Now we are ready to go with our thread-local copy of the messages queue.

```
// add control messages to MIDI packet queue
_addControlMessages(queueCopy);

// send messages
_flushMessages();
```

```
...
void Device::_addControlMessages(std::queue<midi_message> &messages) {
    struct timeval tv;

    while (! messages.empty()) {
        midi_message msg = messages.front();
        messages.pop();

        // figure out when this packet was added
        gettimeofday(&tv, NULL);
        double elapsedTime = (tv.tv_sec - msg.timestamp.tv_sec) *
1000.0;      // sec to ms
        elapsedTime += (tv.tv_usec - msg.timestamp.tv_usec) / 1000.0;
// us to ms
```

This section is important for our purpose of ensuring that our MIDI controller is as responsive as we think it is, and also for dealing with any old messages. When the message got added to the queue, it was time stamped so that we can see how much time has elapsed later on when we reach this point and are about to transmit it. This is accomplished by getting the current system time via `gettimeofday()`, which returns a `struct timeval`. Here, we compute the difference between the message's timestamp and the current time, and convert the seconds and microseconds components of the `timeval` into milliseconds, which are an appropriate scale for measuring our latency.

Should some horrible freak accident have taken place and delayed our output, the last thing we want is to send old control messages. Someone doing a live performance wants the current values of their instrument parameters to reflect the present and not a bunch of milliseconds ago, so it is safer to just drop any old messages and wait for the next update. We will define an "old" message as an arbitrary 2 ms.

```
        if (elapsedTime > 2) {
            cerr << "Warning, MIDI control message latency of " <<
elapsedTime << "ms detected.\n";
            continue; // drop message
        }
```

Assuming we have not bailed on this pass through the loop, we then add our message to the list of MIDI packets to be sent, re-using the `MIDIPacketListAdd()` code from the earlier version of our program:

```
_addControlPacket(msg.control_index, msg.control_value);
```

Finally, we have the lowly, blocking `_flushMessages()`, which performs the actual work of sending our MIDI packet list.

```
OSStatus Device::_flushMessages() {
    OSStatus res = MIDIReceived(deviceEndpoint, midiPacketList);
    initPacketList();
    return res;
}
```

You now have the tools needed to create robust, multithreaded Leap applications that can separate gesture input from program output and can take full advantage of modern multicore systems. This same structure of having a channel protected by a mutex that is used to pass tasks from one thread to another can be applied to any sort of problem you may encounter as you end up using APIs that may block.

Summary

Multithreaded C++ applications are inherently nontrivial and full of debugging pitfalls. Your best strategy is to abstract the communication between your threads to a simple and well-tested mechanism for safe transportation of messages between your threads. Note that the callbacks used by the Leap library are invoked in separate threads, so take care while sharing data between your callbacks and any other threads your program may be running. This makes it possible for applications to easily receive updates concurrently with the rest of the program's execution, but you are still responsible for appropriately handling any other blocking that may take place in your program.

3

Actual Gestures

When we constructed our MIDI controller, it was based on a primitive but extensible gesture recognition system, into which we could build recognizers of any complexity. In this chapter, we'll discuss both general suggestions for creating gesture-oriented computer interfaces as well as implementation of the built-in high-level Leap gesture components in the SDK.

Computer input

Computers are marvelous. They can do incredibly boring, repetitive tasks such as arithmetic and transfers of data at unbelievable speeds and they can automate all sorts of useful tasks. None of this does any good to anyone, though, without some sort of input and output system, in a way that a car with no doors or a toaster with no bread slots would be useless in the extreme.

The ways in which humans can interact with modern electrical computers have evolved considerably, from the days of mechanical typewriter-based outputs and punch cards to more modern keyboards, mice, and multi-touch input systems with graphical interfaces. With the Leap Motion, we can add a new dimension of input, both literally and figuratively, but we need to decide how to let users issue commands via hand motions and how to have our software translate the gestures into useful input and respond accordingly.

This task is somewhat complicated with a lack of a shared understanding or vocabulary among end-users of gesture-enabled software. There are a few conventions; smartphones, tablets, and modern laptop track pads have popularized some simple interactions such as tapping to represent a click, pressing and sliding to pan, and pinching to scale content. Because using the controller is a new way of communicating with software, we cannot always assume that the person sitting in front of a computer with gesture-enabled programs will know how to direct it, which makes our task a bit more complicated. The likelihood of success of an interface is therefore based on how naturally someone can interact with it. Abstract gestures are far less intuitive, harder to remember, and train people on, as compared to using familiar gestures such as grasping, pointing, swiping, and poking to interact with a skeuomorphic (imitating real-life) interface.

Natural gestures

The most natural modes of gestures are sadly infeasible as far as normal software is concerned. Humans express themselves best when they are engaged in multimodal communication, as in speaking aided by gestures and facial expressions. There has been quite a bit of progress made in speech recognition and detection of emotions, but making use of the gestural information that people convey in everyday speech requires some creative interpretation and understanding of human languages that is beyond the reach of applications today.

A similar problem arises with using sign language, another very natural human-oriented mode of gestural communication. Signing, like speech, is multimodal; mouthing words along with the hand signs in order to accurately and expressively impart meaning to the symbols used to express one's thoughts. It is, likewise, not terribly useful for directing computers due to the fact that it requires a great deal of contextual information and suffers from the same difficulties that arise from any other sort of natural language processing. Not to mention most users would be rather disinclined to learn a signing language in order to give input to their computer.

Given the complexity of interpreting natural and multimodal interactions, we should opt for something simpler. Regressing back to an early stage of childhood development, we can make use of deictic gestures, which are gestures that carry a notion of context. It helps to imagine your users as a young infant, pointing at what they want, trying to convey meaning without the ability to form complex sentences. This is, of course, not far from the mental state of the typical end-user.

The context in which our gestures are designed is the software that the user is trying to interact with. When they want to scroll a page, or zoom into an image, or fling something away by swiping, they are naturally making indications within the virtual environment of your program. Think of a typical smartphone photo gallery; even though you are dealing with an abstract collection of digital images they are spatially arranged in a carousel. When you flip through the carousel, the images go to a place in the virtual space created by the photo album software, either being moved to the left or the right. This imaginary construct of space lends intuitiveness to the swiping gesture. If you fling a photo to the left and it vanishes, you can retrieve it by swiping back to the right, moving the photo through space the same way you would move a physical object.

Try to imagine a virtual environment and the actions your users can perform on objects in that space when designing a gesture interface. As an example, consider a gesture-enabled web browser. A good visual metaphor could be placing the page history in a stack of web pages, allowing the user to use a swiping or grabbing gesture to pop items off in order to go back in the history. This would be a good example of creating a simple virtual space that helps the user relate to the context of their actions, and as a result, we can use an intuitive gesture to manipulate that environment. For comparison, a far inferior interface would be an unenhanced browser with an arbitrary gesture, say, moving your finger in an L shape (used by some mouse-based browser plugins), to go back through the history. Not only is the gesture going to be impossible for a new user to guess, but it has no relation to the object or environment they are trying to manipulate.

> *"We hope that people will model interaction based off physical norms, in which case things are much more fluid and analog than binary as gestures often are and computing has been in the past."*
>
> - David Holz, co-founder of Leap Motion

Our goal should be to eliminate any proxies or indirect interactions and instead let the user interface with an application as naturally as they would with real, physical objects in space. Very few users these days are going to take the time to read an instruction booklet or memorize abstract gestures. Instead they will want to use their considerable experience in using their hands to manipulate things deictically, even if the world in which this takes place is on a computer screen.

 Slight tweaks to real-world environments can greatly enhance the ability of users to interact with them. One example is the Leap DJ software, which simulates a table with two turntables and a DJ mixer. They discovered that the controls at the back of the table were difficult to see and reach, but by slightly inclining the table forward, the controls were much easier to touch in space.

Our challenge now is to connect the natural motions of hands to our software, preferably, at a high level without getting too bogged down in the details. Practically, gestures can be quite tricky to define and detect, and whenever we invent new ones the discoverability of our interface decreases. How thoughtful it is then, that the Leap SDK is stocked with a handy collection of easily configurable and usable gesture recognizers! Let's take a look then, shall we?

Receiving gestures

In order to receive information and updates about gestures that the controller can recognize for us, we must first politely inform our trusty controller of our desires in the onInit callback:

```
void Listener::onInit(const Leap::Controller &controller) {
    controller.enableGesture(Leap::Gesture::TYPE_SCREEN_TAP);
    controller.enableGesture(Leap::Gesture::TYPE_SWIPE);
}
```

And thus for the remainder of the controller's existence it shall diligently try its utmost to notify us when it believes someone has swiped their fingers through the air in a deliberate fashion.

How are we to be notified? When our onFrame callback retrieves a Frame we can ask it if there are any gestures taking place in that frame. Leap::Frame::gestures() returns a Leap::GestureList that contains a list of generic Leap::Gesture instances. A type of Leap::Gesture can be determined, and a suitable subclass with type-specific augmentations can then be constructed:

```
Leap::GestureList gestures = controller.frame().gestures();
    for (Leap::GestureList::const_iterator it = gestures.begin(); it
!= gestures.end(); ++it) {
        Leap::Gesture gesture = *it;
        switch (gesture.type()) {
```

Here we can retrieve the gestures that were picked up in the latest frame, and loop through them. Each pass through the loop, we are given a base Leap::Gesture instance, which can be queried to determine its type. Once we know the proper subtype we can then instantiate a gesture subtype instance using its constructor, passing in the base instance:

```
case Leap::Gesture::TYPE_SCREEN_TAP: {
    Leap::ScreenTapGesture tap = Leap::ScreenTapGesture(gesture);
```

Gesture subtypes can contain additional information that is relevant to that particular gesture. `CircleGestureshave` `.center()` and `.radius()`, `ScreenTapGestures` have `.position()`, and so on. They also inherit a few common accessors from the `Leap::Gesture` base class that can be queried for the current state of the gesture and the hands or `pointables` that are performing the gesture. Also worth noting is the fact that our case statement is wrapped in curly braces, which allows us to declare variables like `tap` inside the scope of the `case` statement.

Some gestures are ongoing, such as when the user draws a circle. When the controller first decides that a circle gesture has been started, the next frame will contain a gesture with `state()` of `STATE_START` and as long as a circular motion is being followed by the `pointables()` each subsequent frame will have gestures in the `STATE_UPDATE` state, finally followed by the `STATE_STOP` state.

It is worth noting that the precise meanings of the gesture states differ between subtypes; circles can keep updating for a long time since it will allow the user to keep making circular motions for as long as they wish. In contrast, the tapping gestures are triggered only once and always with the state `STATE_STOP`, and `progress()` will always return `1.0`.

The settings for recognizing and triggering the various gesture subtypes can be adjusted to meet our needs or tastes. Personally, I find the screen tap gesture a bit tricky to trigger, so why not sacrifice a very little bit of latency for improved detection accuracy? Looking in the reference documentation for the `ScreenTapGesture`, we see that the default moving window for detection is `0.1` seconds. If our application is more concerned about accuracy than speed of recognition (in contrast to our MIDI controller application) then we can tweak the parameters of the recognition engine to suit our needs by accessing the configuration of our controller:

```
    Leap::Config config = controller.config();
    bool setConfigSuccess = config.setFloat("Gesture.ScreenTap.
HistorySeconds", 0.3);
    config.save();
```

Here, we increase the detection window from `0.1` seconds (the default) to `0.3` seconds, allowing more time for the recognizer to confirm that a tap has occurred, reducing the chance of a false positive or negative.

WindowFlinger – a high-level gesture application

Ok, yeah, that's all cool and stuff. The real question is: how do we make some cool shit with this? Glad you asked.

We could spend time constructing some contrived cookbook recipe browser interface or something, but here's a sweeter idea: let's hook up our controller to toss operating system windows around.

Many popular operating systems these days include a graphical interface based on the decades-old work, famously pioneered at the Xerox Palo Alto Research Center, utilizing windows, icons, mice, and pointers, derided as the "WIMP" interface by serious nerds. This has served most users quite nicely for a long time it is true, but let's see if we can do better and let people use their controller to manage their system windowing interface.

Application windows were popularized in the Macintosh MultiFinder in 1987 and provided a friendly graphical interface consisting of containers for views of a given application. They could even be properly layered on top of each other using some Ingenious Optimizations, which was a pretty cool trick that took Microsoft several years to replicate in windows. These windows came with title bars which a user could grab with the mouse and drag it to where they wanted to put it, and resize areas that could be used to expand or shrink the size of the window, allowing multiple windows to be visible at the same time. As display resolutions have increased, along with the ability for modern computers to run dozens of applications at once, managing and arranging windows can be somewhat daunting at times, and certainly not as fun as it could be with a proper interface to take advantage of gesture interactivity.

The virtual environment we should consider is the layers of windows, along with their dimensions and positions in space. One could even consider layering as a representation of windows' depth. For our purposes, the title bar is no longer necessary; we should allow the user to point where they want the window to be, keeping in mind the principle of imagining our end users as possessing an infant's level of communication and mental ability. The Motion API can indicate to us if the user is trying to grow or shrink a window. The high-level `Leap::Gesture` classes can be used to select a window with a screen tapping gesture (`ScreenTapGesture`) and to dock a window in a segment of the screen with a swiping motion (`SwipeGesture`). The screen intersection should serve nicely to let the user relocate windows by pointing at the desired location.

These operations are generic enough to work with windowing systems on different platforms, so we should abstract the platform-specific functions into a driver that can be implemented for different systems. It should handle maintaining a reference to a window, repositioning and resizing:

```
namespace flinger {
typedef void* flingerWinRef;
```

Our application will need a way to reference windows, but it won't know what underlying type is used by the operating system. Nor should it care.

```
class Driver {
public:
    virtual ~Driver() {};

    virtual void releaseWinRef(flingerWinRef win) = 0;
```

We'll need a way to free our reference to a window, along with a way to obtain one:

```
    virtual const flingerWinRef getWindowAt(double x, double y) = 0;
```

To let the user select a window, we should have a way to determine what window they are trying to select based on the screen coordinates. x and y are doubles because that's what `Leap::Screen::intersect()` vector component types are, in case you were wondering:

```
virtual Leap::Vector getWindowSize(const flingerWinRef win) = 0;
virtual const Leap::Vector getWindowPosition(const flingerWinRef win)
= 0;
    virtual void setWindowPosition(const flingerWinRef win,
Leap::Vector &pos) = 0;
    virtual void setWindowCenter(const flingerWinRef win, Leap::Vector
&pos) = 0;
    virtual void setWindowSize(const flingerWinRef win, Leap::Vector
&size) = 0;
    virtual void scaleWindow(const flingerWinRef win, double dx,
double dy) = 0;
```

The rest of these are some basic utility setters and accessors for translating and scaling a window, and should be self-explanatory. We'll use Vectors for representing coordinates, ignoring the Z coordinate for now.

In our controller, we should request to be notified of swipe and tap motions for window docking and selection:

```
void Listener::onInit(const Leap::Controller &controller) {
    controller.enableGesture(Leap::Gesture::TYPE_SCREEN_TAP);
    controller.enableGesture(Leap::Gesture::TYPE_SWIPE);
```

First order of business is to keep a track of which window is currently selected. We can stash that in an instance variable:

```
protected:
    Driver *driver;
    flingerWinRef currentWin = NULL;
```

Now, on to the frame callback, we want to let the user tap the window they wish to select. Let's check for a tap gesture and the screen being tapped:

```
void Listener::onFrame(const Leap::Controller &controller) {
    Leap::ScreenList screens = controller.calibratedScreens();
    Leap::GestureList gestures = latestFrame.gestures();
```

Iterate through our gesture list and handle them:

```
    for (Leap::GestureList::const_iterator it = gestures.begin(); it
!= gestures.end(); ++it) {
        Leap::Gesture gesture = *it;
```

Next, we check the type of the gesture:

```
        switch (gesture.type()) {
            case Leap::Gesture::TYPE_SCREEN_TAP: {
                Leap::ScreenTapGesture tap = Leap::ScreenTapGesture(g
esture);
```

That was a lot of effort but it was totally worth it, because now we've got a sweet tap gesture variable. What's so sweet about it? Well `ScreenTapGesture` happens to contain a `pointable()` method which tells us which pointable object performed the tap. With the combination of our `ScreenList` and `Pointable` we can find the nearest screen that the user was pointing at when they tapped. Here's a handy function that returns a vector describing the screen coordinates where a Pointable is pointing, or an empty, invalid vector if it cannot determine a ray intersection:

```
// where is this pointable pointing (on the screen, not in space)?
static Leap::Vector pointableScreenPos(const Leap::Pointable
&pointable, const Leap::ScreenList &screens) {
    // need to have screen info
    if (screens.empty())
        return Leap::Vector();

    // get screen associated with gesture
    Leap::Screen screen = screens.closestScreenHit(pointable);
    if (! screen.isValid())
        return Leap::Vector();
```

Here we make use of the handy `ScreenList.closestScreenHit()` method, which we can use to find what screen is being pointed at. If the pointable is parallel to or pointed away from all screens, then we'll get an invalid screen back.

```
// get point location
Leap::Vector cursorLoc = screen.intersect(pointable, true);
if (! cursorLoc.isValid())
    return Leap::Vector();
```

Here's our old friend `Screen.intersect()`. We'll ask for the screen-normalized coordinates of the pointable.

```
double screenX = cursorLoc.x * screen.widthPixels();
double screenY = (1.0 - cursorLoc.y) * screen.heightPixels();
return Leap::Vector(screenX, screenY, 0);
```

And here we return the screen pixel coordinates, with Y values increasing from the top down to the bottom.

Window management abstraction

At this point in the program we now know that the user performed a tap and we know exactly which screen coordinates they were pointing at when tapping. What we want to do now is to see if a window is positioned at the screen coordinates of the pointable ray intersection. Our driver will take care of the platform-specific details.

```
flingerWinRef win = driver->getWindowAt(screenLoc.x,
screenLoc.y);
    if (win == NULL)
        continue;

    currentWin = win;
```

The job of `getWindowAt(x, y)` is to query the window manager and determine if the point falls within the bounds of any windows onscreen, and if so, which window is frontmost. If a window is located, an opaque reference to it is returned and we can stash it away later, in order to move and resize the selected window when we receive the appropriate hand motions.

Firstly, we need a simple and intuitive interface for our user to indicate where they would like the window to sit in the virtual environment generated by the window manager. While personally I think it would be nifty to be able to grab a window to drag it, unfortunately the controller does not do particularly well with clenched fists, although doubtless this will be perfected in a future update.

As described earlier, translating the screen via a pointable ray should suffice for our purposes. Before handling that case it would be good to make sure that the user has one finger or pointable extended, and a window already selected via a screen tap:

```
if (currentWin && latestFrame.pointables().count() == 1) {
```

Now, with a single function call we can retrieve the current screen coordinates of our pointable ray intersection. Nothing like some code reuse to brighten up one's day:

```
Leap::Vector hitPoint = pointableScreenPos(latestFrame.pointables()
[0], screens);
```

I'll admit the `Leap::Vector` type in the preceding code seems a little cumbersome, especially if we decide to change it to a more suitable type for two-dimensional graphics. Replacing it with a `typedef`, an `auto` type, might be a decent idea here. It's really a matter of personal preference.

Anyway, if we've got a spot to move the window to, we can delegate the messy business to our driver:

```
if (hitPoint.isValid())
    driver->setWindowCenter(currentWin, hitPoint);
```

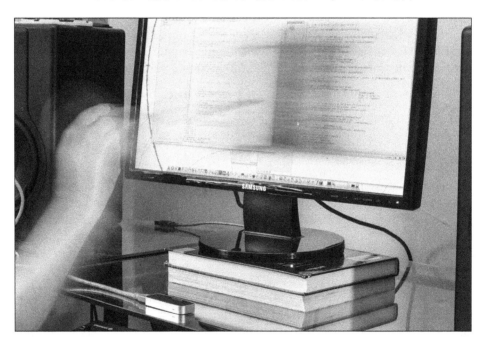

Now, for handling window resizing, we'll leave it up to the Leap SDK to make a guess about whether the user is attempting to indicate a desire to scale the currently selected window.

Naturally it can be a messy business to tell what sort of motions the user is trying to communicate with their hands, so the SDK, in its typical prognostic helpfulness, has just the thing we need here. It contains built-in functions for determining how confident it is that the user is performing a translation, scale or rotation motion, along with a magnitude. One of the benefits of using the motion API components is that as the capabilities and accuracy of the Leap driver are improved, our applications gain the enhancements for free. Also, it's pretty sweet if you're lazy.

We can even easily adjust the sensitivity a bit by deciding what confidence threshold we wish to accept as valid input, with lower confidence cut-offs making it easier to input the gesture but increasing the risk of a false positive.

The one parameter that each of the three sets of motion methods take is a reference frame. What we are really querying is the relative amount of scaling, rotation, or translation that has occurred over a particular period of time, from the reference frame until the present. We can certainly easily retrieve an older frame from our controller to use as a reference frame, though this does add a small amount of variation to the detection window size, depending on the current configuration (precise or rapid processing) and the power of the machine. The more data frames per second processed, the shorter the same interval of frames will be from configuration to configuration.

The same motion API methods are available for both Hand and Frame instances. The simplest usage is as follows:

```
Leap::Frame latestFrame = controller.frame();
Leap::Frame refFrame = controller.frame(handGestureFrameInterval);
if (currentWin && refFrame.isValid()) {
    float scaleProbability = latestFrame.scaleProbability(refFrame);
if (scaleProbability > handGestureMinScaleProbability) {
    float scaleFactor = latestFrame.scaleFactor(refFrame);
    driver->scaleWindow(currentWin, scaleFactor, scaleFactor);
}
}
```

Our reference frame is taken previously from `handGestureFrameInterval` frames. If our controller actually has a frame saved from that long ago, then the returned frame will be valid, and we can query for the probability that the user was attempting to perform a movement of some sort that probably indicated they wanted to scale something. If we have a window currently selected, and the motion detection confidence level is above our threshold constant `handGestureMinScaleProbability`, then we ask the driver to scale the window by the appropriate magnitude. Couldn't be simpler!

As far as appropriate values for the two constants go, some experimentation may be in order. The SDK motion API example code uses 0.40 as the confidence threshold and an interval of one frame for reference, but other values may produce more desirable results, depending on your application and tastes. Ideally you would scale the reference frame interval by the current estimated data frame rate (which the sample application appears to do, but doesn't actually).

Window docking

Now that selection, translation, and scaling of windows is worked out, we can add support for our final supported gesture, flinging the window into a section of the screen to dock it. Imagine the user poking a window and then tossing it in whichever direction they choose with literally a flick of the wrist. Who wouldn't want to have the ability to dismissively toss around windows?

Implementing the gesture detection is straightforward enough; the `Swipe` gesture is triggered when the user draws a straight (-ish) line in the air, rapidly. In addition the gesture can be active while the user is still drawing the swipe, which allows us to provide a little bit of instant visual feedback.

Going back to our previous `switch(gesture.type())` block we can insert another case to handle `Swipe` gestures:

```
case Leap::Gesture::TYPE_SWIPE: {
    auto swipe = Leap::SwipeGesture(gesture);
    auto direction = swipe.direction();
```

The `swipe` direction is a vector representing the movement of the pointable, with the amount of change in each axis specified by the values of each vector component. The actual values are in a very small range so we'll want to multiply each component by a larger value when computing a window translation offset in pixels. When the `Swipe` is ongoing (`STATE_UPDATE`) the currently selected window should be scooted accordingly to provide an immediate visual indication that the user is now in the middle of flinging the window:

```
                            case Leap::Gesture::STATE_UPDATE:
                    if (currentWin) {
                        auto magnified = direction * 100;
                        auto curPos = driver-
>getWindowPosition(currentWin);
                        if (curPos.isValid()) {
                            curPos.x += magnified.x;
                            curPos.y += magnified.y;
                            driver->setWindowPosition(currentWin,
curPos);
                        }
                    }
```

Check out that sweet overloading action for `Vector` with the `*` operator. It multiplies each component by a scalar and returns the scaled `Vector`. Window scooting is then a simple matter of offsetting the current window position by the X and Y components of our magnified `Vector`.

Once the swipe gesture is complete, it is time to stash the window in a segment of the screen corresponding to the swipe direction, either top, bottom, left or right. In order to determine the region in which window can be moved, we must compare the X and Y offsets of the direction vector. If the swiping pointable was moved left, then X will be less than 0, if moved right, then X will be greater than zero. An upward swipe will be indicated by a positive Y value, and a downward one with a negative Y value.

We should expect nonzero values for both X and Y because the controller is sensitive enough to detect minute movements, thus we should make our determination of which axes to respect based on whichever has a greater absolute value. We'll save our decision of the window fling direction for processing after we've iterated over all of the active gestures in the current frame. This will have the effect of only handling one fling at a time even if multiple pointables are outstretched when the fling occurred. It's not a perfect solution but it suits us just fine in this case.

Here is one way of determining the desired window docking location and saving the location in `dockPos`:

```
                            case Leap::Gesture::STATE_STOP:
                    if (direction.x > 0 && direction.x >
direction.y)
                        dockPos = FLINGER_DOCK_RIGHT;
                    else if (direction.x < 0 && direction.x <
direction.y)
                        dockPos = FLINGER_DOCK_LEFT;
                    else if (direction.y < 0 && direction.y <
direction.x)
                        dockPos = FLINGER_DOCK_BOTTOM;
                    else if (direction.y > direction.x)
```

```
                dockPos = FLINGER_DOCK_TOP;
        break;
```

We handle this at the end of the swiping movement, reported via STATE_STOP. The swipe end is triggered when the current pointable velocity falls below a certain threshold.

When dockPos is set, it's time to size and position the currently selected window into that region. We'll set the width or height to one half of the current screen's width or height and move it to the proper edge:

For simplicity's sake we will assume we are using the first available screen in our ScreenList. At the end of this segment, we can deselect the current window, so that it stays docked and doesn't get translated or scaled out of its little corner.

```
    if (currentWin && dockPos != FLINGER_DOCK_NONE && ! screens.
isEmpty()) {
        auto dockScreen = screens[0];
        auto width = dockScreen.widthPixels();
        auto height = dockScreen.heightPixels();

        float destX, destY, destWidth = 0, destHeight = 0;
        switch (dockPos) {
            case FLINGER_DOCK_LEFT:
```

```
                destX = 0;
                destY = 0;
                destWidth = width / 2;
                destHeight = height;
                break;
            case FLINGER_DOCK_RIGHT:
                destX = width / 2;
                destY = 0;
                destWidth = width / 2;
                destHeight = height;
                break;
            case FLINGER_DOCK_TOP:
                destX = 0;
                destY = 0;
                destWidth = width;
                destHeight = height / 2;
                break;
            case FLINGER_DOCK_BOTTOM:
                destX = 0;
                destY = height / 2;
                destWidth = width;
                destHeight = height / 2;
                break;
            default:
                cerr << "Got dockPos set to unknown enum value\n";
                break;
        }

        if (destWidth && destHeight) {
            Leap::Vector pos = Leap::Vector(destX, destY, 0);
            driver->setWindowPosition(currentWin, pos);
            Leap::Vector size = Leap::Vector(destWidth, destHeight,
    0);
            driver->setWindowSize(currentWin, size);

            setCurrentWin(NULL);
            return;
        }
    }
```

As seen here, docking the window is a matter of fetching the current width and height using `Screen.widthPixels()` and `Screen.heightPixels()`, and then repositioning the window so that it is flush with the docking side and scaling it to fill up that half of the screen. Like before, our platform-independent application operates on the boundaries of our virtual environment, and we let the OS driver take care of the messy business unrelated to gesturing. This separation allows for a cleaner application code structure as well as reusability of the bulk of our application on different platforms. We also utilize the Leap SDK's interface to query the screen dimensions, further increasing the portability of our library.

Driver implementation – Mac OS X

Details of the Mac OS X window flinger driver are provided here for completeness. It does not relate to the Leap Motion *per se*, but it does provide a useful and interesting illustration of the layer of platform-specific details and implementation. Skip ahead if you don't care about the gruesome details of manipulating OS X windows.

Accessibility API

Regrettably, OS X does not provide an API for directly accessing the attributes of windows not owned by the current application, which sort of throws a wrench into our plan to fling windows. It does, however, furnish developers with a library for generic property setters and accessors of UI elements for the purposes of accessibility, enabling applications to provide alternative methods of interface feedback for vision- or hearing-impaired users. Examples of this could be an application that speaks the current window title, or changing button labels to be large and high-contrast.

The AX API, as it is known, allows us to retrieve and change the dimensions and coordinates of any window on screen, as long as we have a reference to it. The other piece of the puzzle is getting a list of visible windows and getting a handle to the top-most location of the user's window selection screen tap gesture. This capability is partially provided natively with the Core Graphics API's `CGWindowListCopyWindowInfo()` function.

There is a catch to our scheme though; the AX API is not actually enabled by default. There are two ways to enable it: either it can be manually activated in the Universal Access system preference, or our application can enable it if sufficiently authorized.

Normally, it would be possible to use the Authorization API to elevate the privileges of our application to a sufficient level to make the call to `AXMakeProcessTrusted()` but we hit a bit of a snag here. As it happens, applications that have loaded a shared library are prevented from full system access for security reasons; mostly to prevent hijacking by a shady dylib in the dynamic library search path or environment.

Given this sordid state of affairs, let's settle for manually enabling the AX API:

flinger::MacDriver

The interface for our Mac flinger driver is as follows:

```
#include <CoreFoundation/CoreFoundation.h>
#include <CoreGraphics/CoreGraphics.h>
#include <ApplicationServices/ApplicationServices.h>
#include "FlingerDriver.h"

namespace flinger {
```

```
class MacDriver : public Driver {
public:
    MacDriver();
    virtual ~MacDriver() {};

    virtual void releaseWinRef(flingerWinRef win);

    virtual Leap::Vector getWindowSize(const flingerWinRef win);
    virtual const flingerWinRef getWindowAt(double x, double y);
    virtual const Leap::Vector getWindowPosition(const flingerWinRef
win);
    virtual void setWindowPosition(const flingerWinRef win,
Leap::Vector &pos);
    virtual void setWindowCenter(const flingerWinRef win, Leap::Vector
&pos);
    virtual void setWindowSize(const flingerWinRef win, Leap::Vector
&size);
    virtual void scaleWindow(const flingerWinRef win, double dx,
double dy);

protected:
protected:
    // passed to CGWindowListCopyWindowInfo
    // to control which windows we are testing
    CGWindowListOption listOptions;

    // last retrieved set of window information
    CFArrayRef windowList;

    // update windowList with current window layer attrs
    void updateWindowInfo();

    // search for a window containing (x,y) at depth winIdx
    // belonging to process pid
    const flingerWinRef findWindowForPID(int pid, int winIdx, double
x, double y);

    CGSize _getWindowSize(const flingerWinRef win);
    CGPoint _getWindowPosition(const flingerWinRef win);
  };

}
```

The public methods for our driver are defining our implementation of the pure virtual methods in the `Driver` base class. The protected methods and data will be used internally to avoid repetition and to cache information retrieved from CoreGraphics and AX.

In our driver setup, we should check and see if the AX API is enabled, and emit an error if not. We'll use the profoundly unfriendly `stderr` file descriptor to indicate why our application failed to start, because we hate our users.

```
MacDriver::MacDriver() {
    if (! AXAPIEnabled()) {
        // try elevating privs to request AX access
        // this is broken because we can't launch with
        // AX setgid privs and load the libLeap dylib
        // http://lists.apple.com/archives/accessibility-dev/2009/Oct/
msg00014.html
        cerr << "Please enable the accessibility API" << endl;
        exit(1);
    }

    windowList = NULL;

    listOptions = kCGWindowListExcludeDesktopElements;
    listOptions |= kCGWindowListOptionOnScreenOnly;
}
```

Here we initialize our options for window layer info call to return only normal, visible application windows. We don't care about moving things like the dock, or minimized around. The call to query these windows and store the result is as follows:

```
void MacDriver::updateWindowInfo() {
    if (windowList)
        CFRelease(windowList);
    windowList = CGWindowListCopyWindowInfo(listOptions,
kCGNullWindowID);
}
```

The second parameter to `CGWindowListCopyWindowInfo` allows us to specify another window as a reference point. We have no need for this feature, as we want to get a view of all available windows. It returns a core foundation array reference (`CFArrayRef`) of `CFDictionaryRef`, containing attributes for each window. We're responsible for freeing the array reference when we're finished with it, so we release any existing reference before updating.

On to our first public driver method, `const flingerWinRef MacDriver::getWind owAt(double x, double y)`. The purpose of this method is to return a reference to the top-most window containing the specified coordinates. This can be achieved by iterating over each window in `windowList` until we locate one that contains the target point. `windowList` already contains a list of processes with windows ordered front to back, so that's handy.

Note that we have a list of processes with visible windows, not a list of windows. We can get a list of windows for a given process ordered front to back; so each time we encounter a given process while iterating, we should assume that the next window in the process list is what is being referred to by `windowList`. This layout of windows, by depth, is somewhat confusing and not documented properly, but with some experimentation it was possible to come up with a solution of locating the topmost window containing a given point:

```
const flingerWinRef MacDriver::getWindowAt(double x, double y) {
    // get latest window list
    updateWindowInfo();

    flingerWinRef ret = NULL;

    // map of process ID to last checked application window index
    map<int,int> checkedProcWindows;

    // windowList contains a list of information about each window on-
screen
    // unfortunately it does not contain the AXUIElementRef which we
need to
    // get and set the window bounds and position.
    // each iteration we must determine which process has the next
window in
    // the list, and check the front-most unchecked window for that
process.
    for (int i = 0; i < CFArrayGetCount(windowList); i++) {
        // check each application's windows in turn. windows appear to
be ordered by "depth" already

        CFDictionaryRef info = (CFDictionaryRef)CFArrayGetValueAtIndex
(windowList, i);

        // get pid
        int pid;
        CFNumberRef pidRef = (CFNumberRef)CFDictionaryGetValue(info,
kCGWindowOwnerPID);
        CFNumberGetValue(pidRef, kCFNumberIntType, &pid);
```

```
        if (! pid) continue;

        // which window did we last check for this process? check the
next one in the stack
        int procWinIdx;
        if (checkedProcWindows.find(pid) != checkedProcWindows.end())
            procWinIdx = checkedProcWindows[pid] + 1;
        else
            procWinIdx = 0;
        checkedProcWindows[pid] = procWinIdx;

        // get layer
        int layer;
        CFNumberRef layerRef = (CFNumberRef)CFDictionaryGetValue(info,
kCGWindowLayer);
        CFNumberGetValue(layerRef, kCFNumberIntType, &layer);

        // would be nice to be able to select windows from system
layers,
        // but they always appear to be on top
        if (layer != 0) continue; // normal application layer

        // search for window owned by process pid with a frame that
includes the point (x,y)
        cout << "Checking point in " << x << "," << y << endl;
        ret = findWindowForPID(pid, procWinIdx, x, y);

        // found our window?
        if (ret) break;
    }

    return ret;
}
```

Firstly, `windowList` is updated to contain an array of dictionaries describing the visible application windows on screen in order of visibility. Each window is checked in turn with `findWindowForPID` to see if it contains the point passed in. We also scope out the window's layer ID, which indicates to us whether or not it is a normal, moveable application window.

In `getWindowAt`, we don't have much in the way of a useful window reference that can be used by the accessibility API to get position and size information, but we do have the process ID of the window owner. Because we can also get the lists of windows (including useful references) for a given PID, it is just a matter of checking the next window in the application's stack, each time we encounter the same PID to see if it contains our hit point.

Getting the position and size attributes of a window that doesn't belong to our application begins with getting an AXUIElementRef for the window in question from the application's process ID and window depth index. With the accessibility user interface element reference, we can retrieve size and position attributes. Because the accessibility UI is very generic and is meant to represent all sorts of elements, the types that it deals in are also rather generic and abstract. What we need to do is to convert the more abstract types returned from the AX API into more specific, native representations such as CGPoint and CGSize. The following code for findWindowForPID shows how to accomplish this. Note that a great deal of error checking is omitted for clarity.

```
const flingerWinRef MacDriver::findWindowForPID(int pid, int winIdx,
double x, double y) {
    flingerWinRef ret = NULL;

    // get all windows for this application
    AXUIElementRef appRef = AXUIElementCreateApplication(pid);

    // get app windows for appRef
    CFArrayRef appWindowsRef;
    AXUIElementCopyAttributeValue(appRef, kAXWindowsAttribute,
(CFTypeRef *)&appWindowsRef);

    // check next window in the stack, which we should find at winIdx
    long winCount = CFArrayGetCount(appWindowsRef);
    if (winIdx >= winCount) {
        CFRelease(appWindowsRef);
        return ret;
    }

    // get a reference to the window at index winIdx
    AXUIElementRef winRef = (AXUIElementRef)CFArrayGetValueAtIndex(app
WindowsRef, winIdx);
    CFRetain(winRef);
    CFRelease(appWindowsRef);

    // now that we've got our proper window reference for AXUI, we can
ask about
    // window size and position
    CFTypeRef encodedSize, encodedPosition;
    CGSize size; CGPoint position;
    // get size/position as opaque references
    AXUIElementCopyAttributeValue(winRef, kAXSizeAttribute, (CFTypeRef
*)&encodedSize);
```

```
    AXUIElementCopyAttributeValue(winRef, kAXPositionAttribute,
(CFTypeRef *)&encodedPosition);
    // convert references into CGSize and CGPoint types
    AXValueGetValue((AXValueRef)encodedSize, kAXValueCGSizeType, (void
*)&size);
    AXValueGetValue((AXValueRef)encodedPosition, kAXValueCGPointType,
&position);
    CFRelease(encodedSize);
    CFRelease(encodedPosition);

    // check to see if this window contains our test point
    CGPoint testPoint = CGPointMake(x, y);
    CGRect winRect = CGRectMake(position.x, position.y, size.width,
size.height);
    if (CGRectContainsPoint(winRect, testPoint)) {
        ret = (void *)CFRetain(winRef);
    }

    CFRelease(winRef);
    return ret;
}
```

Given a process ID, a window index and a point, this method demonstrates how we can see if the specified window contains the point. We request the application's window references by passing kAXWindowsAttribute to AXUIElementCopyAttributeValue, which returns a CFArrayRef of AXUIElementRefs. We pick out the next window in the stack, provided to us via winIdx, and then use AXUIElementCopyAttributeValue a couple more times to fetch kAXSizeAttribute and kAXPositionAttribute. AXValueGetValue performs the useful task of converting an abstract type reference to a more specific, useful, and concrete OS type; for example, used here to convert the encodedPositionCFTypeRef* to a CGPoint type specified by kAXValueCGSizeType.

The rest of the driver method implementations follow a similar pattern of using AXUIElementCopyAttributeValue and related functions to get and set window position and size parameters for window references, which are exposed as flingerWinRefs.

The complete, functional application code can be found at https://github.com/revmischa/leap-windowflinger.

Summary

Thanks to our driver abstraction layer and opaque window references, we can write our code that deals with the controller and performs platform-specific manipulations, and yet is set up to be ported to other operating systems easily, just by implementing the proper driver methods and specifying which source files to compile.

A number of useful modules for creating portable Leap applications with few dependencies are additionally provided by the SDK in the `util` directory. `LeapUtil.h` contains a limited but practical set of math-related routines for motion and geometry, smart pointer generics, and a generic camera class that can be used with any sort of 3D. In addition, `engine.LeapScene.h` has utilities for tracking visible objects and geometric primitives in a virtual environment, and has convenience methods for computing pointable ray intersections with scene objects, as well as tracking interactions such as object selection and transformations. Lastly, `LeapUtilGL.h` gives us a small set of primitive drawing methods for OpenGL, such as arrows, cylinders, grid planes, and axes. It also has a very simple but helpful cross-platform include statement for locating the proper OpenGL utility header (`glu.h`), which is convenient. While these utilities were likely created primarily to be used internally by the example programs included in the Leap SDK, it's worth noting that the Leap Motion developers were kind enough to comment on and expose their interfaces and code here. Even though these utilities are not strictly related to the Leap functionality, they may be helpful in creating portable application prototypes.

4

Leap and the Web

Compiled Leap applications written in C++ are extremely powerful and very easy to distribute via the Leap Motion app store (Airspace). The controller's usefulness extends beyond native applications though; it can also seamlessly integrate with client-side web technologies.

In this chapter we'll cover the basics of the Leap SDK for the Web, including the following:

- WebSockets
- LeapJS
- Drawing on a canvas

HTML5 and Leap

Creating scripted or compiled desktop applications with the SDK is all well and good. There are plenty of nifty applications which require lower-level interfaces with a particular operating system API, as in our examples of cursor control, MIDI output, and window management. But when it comes to portability, distribution, and simplicity, it is pretty hard to beat the Web and HTML5.

User-agent standards have become massively more robust in recent times, making powerful mechanisms such as WebSockets, two-dimensional and three-dimensional graphics capabilities, and much more available. All that is necessary to tap into these powerful toolkits is a bit of HTML and JavaScript. Whereas in the earlier days of the Web, if one wanted to interface with a system service connected to specialized hardware such as Leap, as a minimum, a browser plugin would be required. With modern web browsers such tomfoolery is no longer required, and enabling Leap support for any old web page is as simple as could possibly be expected given the circumstances.

Old-time web developers may be skeptical of this claim and ask, "How can some plain old JavaScript receive frame data from the controller without any special browser support?" The answer is surprisingly simple and ingenious: the Leap Motion service comes with a WebSocket server that emits frame data to connected clients.

WebSocket

The WebSocket protocol is defined by RFC6455 and the IETF. While it resembles traditional network sockets that can be accessed with browser-based JavaScript, it is not exactly the same. Two-way communication is established with a handshake very similar to an HTTP request. The client sends a **request method** and **path**, **hostname**, **encryption key**, and a **Connection: Upgrade** header field. Once the server responds appropriately, messages can be passed back and forth inside small WebSocket frames. By requiring the other end of the WebSocket to conform to the WebSocket protocol, the dangers of directing web clients to send malicious data of any old sort to any host are mostly mitigated by the need for two-way communication.

The Leap Motion service

The Leap Motion service is an application running as a daemon on Unix-style platforms (Mac OS X and Linux) and as a service on Windows. Its job is to communicate with the controller hardware devices over USB, post-process frames of hand- and scene-tracking information, and then make this data available to API clients in their native formats. When we use the C++ API to create our frame data **Listener**, it is really communicating with the Leap Motion service behind the scenes, which is why the service must already be started by the user for our `Listener` to receive any updates. In the case of our client-side JavaScript code, the frame data is transmitted over a socket, although the details of that communication are still hidden by the consumer of the Leap API.

LeapJS

While we could certainly write our own JavaScript library to use WebSockets to receive the controller frame data from the service's built-in WebSocket server, that would be a colossal waste of our efforts given the fact that there exists such a library already: **LeapJS**.

LeapJS is an open source library that contains routines and interfaces to make it easy to do absolutely anything anyone might want to do, with the controller in the land of JavaScript. Modules are provided for both HTML5 clients as well as Node. js applications, providing Leap support on the frontend and backend. The frame data, once received, is structured to match the C++, Objective-C, and SWIG-based interfaces that we have already studied.

Receiving the controller frame data and making use of it is even simpler than compiling C++ Leap-based applications; all that is needed is a web browser and a text editor. Though before we float down the calm river of client-side Leap development, we should affix the LeapJS outboard motor to our water-faring craft.

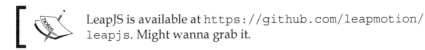

LeapJS is available at `https://github.com/leapmotion/leapjs`. Might wanna grab it.

Let's get back to basics, this time with a little JavaScript. Create a new HTML file in a directory alongside `leap.js` or `leap.min.js` from a clone of the LeapJS repository, and try out this simple demonstration web page:

```html
<html>
    <head>
        <script src="leap.js"></script>
        <script>
            Leap.loop(function(frame) {
                // get all fingers+tools
                var pointables = frame.pointables;
                if (! pointables.length)
                    return;

                for (var i = 0; i < pointables.length; i++) {
                    var pointable = pointables[i];

                    // get fingertip position
                    var pos = pointable.tipPosition;
                    if (! pos)
                        return;
```

```
                        console.log(pos);
                  }
            });
      </script>
   </head>
   <body>
      <h1>Whatup. Peep the JavaScript console.</h1>
   </body>
</html>
```

Assuming that the LeapJS directory is located in the same place as your HTML file, the Leap service is running, and your controller is connected, you should be able to view the coordinates corresponding to your fingertips in your web browser's JavaScript console.

The output of fingertip coordinates in the Chrome developer tools console

Take note of the `Leap.loop()` function; it is one of the ways of receiving updates of tracking data. To use it, pass a callback with one parameter: the received frame. After the callback is finished doing its thing, the Leap client library will call the global browser function `window.requestAnimationFrame()` to chill until the browser is ready to draw the next frame of animation.

`requestAnimationFrame()` is a semi-recent addition to the standard suite of global browser JavaScript functions, designed to make it feasible to create web pages that are well-behaved yet still perform smooth, high frame rate animations, and interactions. It invokes a user-specified callback that should handle the task of painting the next frame, and then the browser performs a repaint operation afterwards. It will execute up to 60 frames per second, though the rate may be slower if there is a lot going on or the page is in a hidden tab or the background.

We do not directly invoke `requestAnimationFrame()` in this case; it is automatically called by the super magical `Leap.loop()` convenience function, along with the WebSocket and `Leap.Controller` setup. It is possible to perform this setup ourselves if we desire more control, but in plenty of cases, all we really want is to do some animation updates whenever the browser is ready to draw a frame.

As in C++ land, we can choose to receive hand tracking data in a callback that is invoked periodically by the Leap API on its own schedule, and we also have the option of polling for data using the `controller.frame()` instance method to fetch a previously recognized frame. By default, `frame()` returns the latest frame, or you can pass in an integer to request an older frame. The controller remembers up to 60 previous frames. While polling or requesting frames from the past, always check `frame.isValid()` to make sure that we've been handed a real frame; if no frame data has been recorded that matches our request, we will receive an invalid frame in response. Many other API calls work in a similar fashion; for example, requesting `Hand` that is no longer in the current frame by ID returns an invalid `Hand`, and the empty returned `Vector` math classes are similar (the `Matrix` classes are returned as an identity matrix). This does eliminate the need to check for null everywhere, but those checks need to be replaced by calls to `isValid()`.

Polling via `controller.frame()` and receiving frame update callbacks from `Leap.loop()` each have their merits and appropriate uses. Polling is useful while already using your own event loop or callback system, such as a main game state loop. It is also handy if you want to receive and process frame updates faster than `requestAnimationFrame()` permits, in the case that frame data arrives more rapidly than browser repaints (pretty much always true). `Leap.loop()` is fantastic for more graphical applications or where simplicity is desired. It behaves well in matters of performance scaling and not using excess CPU cycles when not needed. It is more suited to more visual-based applications where subframe accuracy is not needed and those do not already have a `draw` or `main` loop.

There is a rumored third option for receiving frame updates using a JavaScript version of the `controller.onFrame()` callback, though the lack of documentation may indicate that this is dark magic and not to be invoked, at least at the time of writing this book.

Once we get our hands on an actual `Frame` instance, either by callback or polling, we can do our thing. Our frame contains much of the same data as the C++ and SWIG classes, though it is not quite as featureful, owing to the fact that the C++ API is completely native and the core of the Leap API and SDK, and the JavaScript client is receiving the data that is easily serialized and sent via WebSocket. Performing device-related polling tasks, for example, is not presently possible; there is no JavaScript equivalent of the `Device` class or ability to access `ScreenList`.

The classes exposed in the present version of LeapJS are `Controller`, `Frame`, `Hand`, `Pointable`, the `Vector` and `Matrix` math classes, and the predefined `Gestures`. Work is being done to reach more feature parity with the more native APIs, and it is an open source project should you feel like contributing towards this effort.

JavaScript visualization

Now that we have a hand tracking update callback working nicely, let's do some basic visualization of the data to get a feel for combining the tracking data with some primitive HTML5/JavaScript animation.

HTML5 rules. Why? Everyone has a web browser that will run it more or less identically, without the need for users to use any native installers or download anything. Not that obtaining Leap-enabled native applications is that much more work, the Airspace app store makes that about as easy as one could ask for. From the developer's perspective, using HTML5 and JavaScript greatly simplifies tasks such as drawing graphical primitives, performing network I/O, 3D graphics, and obtaining input. Pretty much everything we need to make any sort of interface is easily accessible and already portable without any need for a time-consuming project and linker setup and dependencies as it most certainly would be if we were to, say, attempt to use OpenGL with C++.

For our next little exploratory LeapJS application, we'll do a little bit of graphics to visualize the current position in space of the tips of any `Pointable` classes found in the current Frame that is being drawn. We'll draw some circles onto the HTML5 `canvas` element's 2-dimensional graphics context (as opposed to the WebGL 3D context) to visualize the fingertip positions.

```html
<html>
    <head>
        <script src="leap.js" type="text/javascript"></script>
        <script type="text/javascript">
            document.addEventListener("DOMContentLoaded", function() {
```

Standard boilerplate HTML here, including the LeapJS library and then setting up our code to execute when the structure of the page is loaded. We want to wait until this event fires so that we have access to all DOM nodes, in particular the canvas into which we want to draw. DOMContentLoaded is an event that is fired after it is safe to start referring to elements referenced in the HTML source, but probably before other elements of the page have been loaded such as images.

For simplicity's sake, we will define our drawing region as having a fixed width and height of 1024 x 768 pixels.

```
var ctx; // canvas 2d drawing context
var w = 1024, h = 768;
var canvas;
Leap.loop(function(frame) {
    // get 2d drawing context for our canvas if
    // it hasn't been set up yet
    if (! ctx) {
        canvas = document.getElementById("drawing");
        ctx = canvas.getContext('2d');
    }
```

getContext('2d') returns a graphics context object that is attached to our canvas that we can call drawing methods on. A context maintains a considerable amount of state, such as the segments of a path being currently drawn, current fill and stroke colors, line widths, and much more. Operations such as stroking or filling the current path are performed on the previously specified state.

Here, we revisit our good friend Leap.loop() to handle our simple frame animation. In the loop, we only retrieve the graphics context if we haven't done so already. Fetching the canvas element and getting the graphics context every single frame would be a totally unnecessary waste of CPU effort. Also, note that variables such as ctx are not global or polluting the rest of our web page's world; they are scoped locally inside our DOMContentLoaded callback.

Speaking of wasted CPU effort…

```
if (! frame.pointables.length) return;
```

This bit is crucial. If there are no pointables detected in our frame, we don't really have anything to do and should not bother with drawing. This cuts down noticeably on CPU load when there are no hands in the frame.

```
// blank out canvas
ctx.clearRect(0, 0, w, h);
```

Erase previously drawn frames. Next, iterate through each hand visible in the current frame:

```
for (var hi = 0; hi < frame.hands.length; hi++) {
    var hand = frame.hands[hi];
```

And then, through each finger or tool attached to that hand, use the `hand.pointables` member.

```
// grab each finger or tool connected to hand
var pointables = hand.pointables;
for (var pi = 0; pi < pointables.length; pi++) {
    var pointable = pointables[pi];
```

We should have a tip coordinate vector for this pointable. We shouldn't always count on it in case of occlusion or other unfavorable visual conditions.

```
// do we know where the tip of the finger or tool is
// located?
var tip = pointable.tipPosition;
if (! tip) return;
```

`tip` is a three-element array containing the x, y, and z coordinates of the finger or tool tip in the Leap coordinate system, in millimeters. We want to translate the x and y coordinates from metaspace into the relative coordinates on the canvas, which we do with some extremely rough and unscientific number mangling. If we were using the C++ API, we could use the `InteractionBox` methods to get normalized coordinates in an absolute range; but we're not, so let's just fake it for now. Note that the y and z coordinates are probably the inverse of what we want to display, so we'll go ahead and negate them. (The graphics context has increasing y values going down the page, but we probably want to have the circles move upwards along with the hand motion.)

```
// get x/y/z coordinates of pointable tips
// and convert to coordinates that roughly
// live inside of the canvas dimensions.
var x = tip[0]*2 + w/2;
var y = -tip[1] + h/2;
```

Good enough for government work, as they say. We'll do the same sort of thing to the depth coordinate, `tip[2]`, but instead of using it as a coordinate, we can approximate the effect by moderating the circle radius by the tip depth.

```
// use depth to control the radius of the circle
```

```
                // being drawn
                var radius = (-tip[2] + 100) / 6;   // random numbers lol
                if (radius < 10) radius = 10;        // not too small!
```

Now we have x and y coordinates and a radius, so we might as well get started drawing our tip circle. In the wacky world of the HTML5 `CanvasRenderingContext2D` interface, as well as many other programmatic graphics APIs, drawing a primitive shape involves defining a path, setting a line thickness and foreground color, and then either stroking (drawing an outline) or filling (if it is an enclosed path, anyway) it with the foreground color.

```
                // begin drawing circle
                ctx.beginPath();
                // centered at (x,y) with radius scaled by depth, in a
                full arc
                ctx.arc(x, y, radius, 0 , 2 * Math.PI, false);
                ctx.lineWidth = 5;
                // color based on which hand it is
                var g = hi % 2 ? 200 : 0;
                ctx.strokeStyle = "rgb(120," + g + ",35)";
                // draw circle
                ctx.stroke();
            }
          }
        });
      });
  </script>
```

The call to `arc()` includes a parameter to specify the arc angle in radians. (2π is a full circle, in case you have forgotten.)

We'll color the circles differently depending on which hand it is, supporting a maximum of two hands. One of the hands will be extra green.

Now that we've got the canvas drawing done, all we need is the actual canvas.

```
    </head>
    <body>
        <canvas id="drawing" width="1024" height="768"></canvas>
    </body>
</html>
```

Summary

And that completes our simple JavaScript visualizer. Let's step back and admire it for a second; we've got a (pseudo-)3D cross-platform visual application available to Leap motion users without any download or install, animating at 60 FPS in a web browser, all in a very modest 50 lines of code or so. When contrasted with all of the difficulties of trying to accomplish the same with C++, the advantages of using LeapJS should be apparent.

Of course, our little visualizer is a rather contrived example that doesn't do much in the way of looking cool or showing off the more advanced interaction features of LeapJS or HTML5. More on that coming right up!

The LeapJS pointable tip visualizer in action

5
HTML5 Antics in 3D

In this final chapter, we'll talk a bit about 3D graphics with the Leap, on the Web. Yes, for reals.

We'll take a look at how to make practical use of an HTML 3D graphics standard called **WebGL**, and a popular compatibility and utility library. Lastly, we'll see how the **Leap** and **Three.js** mesh together. The topics we'll take a look at are:

- WebGL basics
- Three.js abstraction layer and toolkit for 3D
- Combining Leap input with 3D objects

Cross-platform graphics party

Once upon a time there was a software engineer who had a simple desire: he wanted to experiment with crafting a three-dimensional environment, just for the heck of it. Should be simple enough, he thought, after all, there are so many 3D games and widgets out there, it can't be that hard. Everyone uses C++ and OpenGL, right? It's cross-platform, isn't it?

He very quickly began to suspect that things might not go as smoothly as he had imagined at the outset. "It's just `#include <GL/gl.h>`, right?" he asked. "Oh wait, apparently Mac OS X wants `<OpenGL/gl.h>`. And what's this about `glu.h` not being found? The OpenGL tutorial that I found online says to use GLUT, but I don't think that's supported anymore, is it? What's this Linux glx contraption? What on earth is mesa?" he mumbled to himself, getting more desperate with each linker error. "Forget it, I'm joining a monastic order" our hapless developer finally announces, chucking his ThinkPad™ out the window.

Sounds familiar? Believe it or not, creating applications with 3D interfaces doesn't have to be an endless path of suffering and incompatibilities. If you've ever wished, you could just jump right in and start playing with a ready-made OpenGL-style graphical engine, without either falling into the pit of despair of native library support nor giving up the excitement and power of a lower-level API by using some fruity abstract game engine, there is an accessible and ready-made standard, WebGL.

WebGL

Some old-timers may instinctively scoff at the notion of creating a 3D environment in a web browser, using only JavaScript (and maybe some GL shader language). Is it really supported by all modern browsers? Is it accelerated? Does it perform well on mobile devices? Does it have the same capabilities found in a native graphics library that I'd want?

Well, more or less yes. There may be some skeptics out there, especially those who remember the first real attempt at 3D graphics on the web long ago, using virtual reality markup language (VRML). That was an attempt to implement 3D animations with standardized markup rendered via browser plugins, and, quite honestly it was not all that awesome, and we should probably forget it ever existed. Those days are long behind us now, fortunately.

Let's talk about WebGL. It's a well-defined API standard maintained by the Khronos non-profit group, along with other OpenGL-related standards. It is supported by nearly all major browsers, with the usual Redmond-based laggard shambling along with rudimentary, experimental support, to be complete someday soon. WebGL provides access to the GPU, including shader support (rather beyond the scope of this modest guide), and it takes full advantage of the fact it is running inside a web browser by integrating cleanly with the document object model, page elements, canvas container, networking capabilities (such as loading image textures via HTTP, including standard cross-origin request security), and is fully controlled through JavaScript.

What we are really after is to find a way to use 3D graphics in a browser. This capability has existed for some time even without WebGL of course, using software to render a scene into a series of frames composed onto, say, a 2D canvas graphics context. A popular and well-respected JavaScript library, called Three.js, was developed originally to perform this pure software rendering along with a suite of functions for managing cameras, lighting, materials, scene layout, and all other sorts of goodies.

As WebGL emerged on the scene, support for using it for hardware-accelerated rendering was added to Three.js, enabling it to become an extremely handy suite of useful tools and just the perfect level of abstraction, targeted at developers who want to bust out cutting-edge user experiences and take full advantage of modern hardware, without the cruftiness of legacy code and grimy native platform details standing in their way.

This should sound a bit familiar, as the Leap API has the same design. As it happens, by combining Three.js and LeapJS, we gain the ability to rapidly construct 3D environments that are deftly manipulated through gesture input. And we don't have to deal with any linkers or shared object dependency-induced night terrors leading to the renunciation of technology. Not that anything is wrong with becoming a monk, but we're here to talk about the Leap.

Three.js + LeapJS – the awesomesauce

Deciding to use Three.js naturally introduces a dependency into our web client application, but when it comes to 3D programming, it's a very acceptable trade-off. One can think of it as the jQuery for 3D graphics, providing invaluable utilities and shortcuts for common tasks, while greatly enhancing fallback capabilities and smoothing over differences in implementations. LeapJS is naturally pretty useful as well for our purposes.

Let's take a look at combining the two. We can either download the Three.js library and include it locally, or we can link to the latest build hosted by GitHub (or elsewhere). While linking to the latest build is great, because we get the latest optimizations and bug fixes without doing any work at all, downloading the library is usually a better option; development is a tad faster (page reloads are slightly quicker without having to fetch the file from GitHub, especially if you are not geographically proximate to their servers), you can develop offline, and the API is not going to suddenly change on you causing breakage.

 The Three.js project is located at `http://threejs.org`, and the GitHub-hosted build artifact can be linked to at `http://rawgithub.com/mrdoob/three.js/master/build/three.js`.

Assuming that we've got our Leap.js and Three.js files handy, our HTML document will begin with something resembling the following code:

```html
<html>
    <head>
        <style>canvas { width: 100%; height: 100% }</style>
        <script src="leap.js"></script>
        <script src="three.js"></script>
```

No giant surprises here. The `canvas` element style is needed if we want to use the full page for rendering. We aren't actually going to include any `canvas` elements in our HTML, but one is created automatically for us by Three.js later on.

The following program is adapted from the Three.js quickstart with commentary and Leap support added. We can add the program code and `script` tag in either the `head` or the `body`:

```javascript
<script>
    // globals
    var scene, camera, renderer, cube;
    var controller;
```

For simplicity, we'll go ahead and pollute the global namespace with some things we want to have handy for later. We'd want to stash these inside an object, or scope them if we were going to write a real program, or if cap-and-trade JavaScript climate legislation is enacted.

Now, the real fun begins. After doing page setup, we'll create a new, empty Three.js Scenes, which is like a stage to place our objects upon. In Three.js, we can construct objects and manipulate them without them being visible, they'll only appear once added to the `scene`.

```javascript
document.addEventListener("DOMContentLoaded", setup);
function setup() {
    scene = new THREE.Scene();
```

A `scene` acts much like the browser DOM; nodes can have parent and child nodes and properties such as visibility or transformations apply to all children.

Cameras are great. They are an object that can be rotated and translated like any other, but they represent the view that is being drawn on the canvas. A perspective camera has several parameters: **field of view**, aspect ratio, **near clipping** point and **far clipping** point.

The near and far clipping values are the locations of invisible planes that define the area extending outward from the camera that is actually computed for rendering. Anything with a z coordinate falling outside of the range is clipped, that is, hidden.

If you think of the visible region as a cone extending outward from the camera, the field of view would be the angle at the tip of the cone, and the clipping planes truncating the tip and base. Objects inside of the cone's area will be picked up and displayed. Anything outside is not rendered, for efficiency's sake.

The aspect ratio is the ratio of the width to the height of the rendered image, if this does not match the window's aspect, then objects will appear squished or elongated. If we want to allow the user to resize the window while our program is running, then we will need to keep the aspect up-to-date (following code).

Creating a new `PerspectiveCamera` with these parameters is just a constructor away:

```
        var aspectRatio = window.innerWidth / window.
        innerHeight;
    camera = new THREE.PerspectiveCamera(
        35,              // field of view
        aspectRatio, // based on current width/height
                    0.1,            // near clipping point
                    100             // far clipping point
    );
```

Sweet, now we've got a camera. For reference, a common fun thing to do with cameras is to reposition them and then call the `camera.lookAt(position)` method to rotate the camera to point at the position vector.

A camera isn't going to do a whole lot of good unless we have a `Renderer` to go with it, to paint what the camera sees in the form of pixels on a `canvas`. We want a `WebGLRenderer` so that we can take advantage of hardware acceleration, but if we wanted, we could also fall back to a `CanvasRenderer` to perform software rendering with a 2D graphics context if the browser didn't support WebGL.

```
        renderer = new THREE.WebGLRenderer();
        renderer.setSize(window.innerWidth, window.
        innerHeight);
```

Two lines? Not bad for setting up a 3D accelerated rasterizer in a web browser. The astute reader may note that we are setting a fixed width and height in our setup code, and we'll probably want to update this on window resize events.

A nifty feature of doing our graphics in a web page is that we can render into a DOM element that can be manipulated like any other. We can style it, move it around, stick a fancy border on it, append it as a child of another node, and so on. For now, appending the renderer `canvas` to the body will do just fine, and it will automatically stretch to fill the window, thanks to the aforementioned CSS style.

```
document.body.appendChild(renderer.domElement);
```

At this stage, we are ready to render our scene and have it show up on the page. Guess we should probably add something to our scene...

```
var cubeGeometry = new THREE.CubeGeometry(1, 1, 1);
```

The handy utility function `THREE.CubeGeometry` returns a `Geometry` object with the specified width, height, and depth. To render a 3D construct in Three.js, we generally need three things: the geometry, comprising the discrete vertices or otherwise-described shape; we need a material which describes how to color the pixels when the construct is rendered, and a mesh which represents the two combined (making it easy to reuse geometry and materials).

Creating a material and mesh for our cube geometry entails two more function calls:

```
        // create a plain, green material
        var cubeMaterial = new THREE.MeshBasicMaterial({
            color: 0x00AA00
        });
        // combine geometry and material into a renderable
scene object
        cube = new THREE.Mesh(cubeGeometry, cubeMaterial);
```

And now the cube variable contains a green cube mesh. At this point the cube only lives inside a variable, not in our 3D scene that is going to be rendered, so we should add it to the scene. This is a highly complicated procedure involving many complex trigonometric concepts and low-level GPU calls which will be discussed later on. For now just try to follow along:

```
scene.add(cube);
```

Yeah, I was just kidding. Who wants to do things the hard way? Leave that up to the suckers trying to use Java instead.

The last thing we should do for our scene setup is to scoot the camera back a little bit so that it is looking at the scene from a short distance away, instead of chilling in the origin. This is accomplished by changing the camera z coordinate:

```
camera.position.z = 3;
```

Animating rotation

Now, we should start animating and rendering things. In a perfect world we could perform our own `window.requestAnimationFrame()` calls, fetching the latest Leap data frames with `controller.frame()` and repainting the Three.js scene, but at present such a setup is impossible to test due to the incomplete state of the LeapJS library.

Our demonstration integration with Three.js and LeapJS will be simplicity itself; we'll rotate our cute lil' unit cube according to the user's hand rotation. When new hand tracking updates come in from the controller, we'll rotate the cube according to the palm normal of the first returned hand.

A normal is like a ray projecting outward from a surface, representing its orientation. The `palmNormal` property of a `Leap.Hand` is expressed as a unit vector that points outward from the plane described by the palm of the hand. The use of a palm normal for rotating an object works nicely because we can describe the rotation of the cube based on the normal vector without the need for any real math or excessive code, and the result is a very intuitive gesture that mimics the way one rotates an object in the real world.

It is worth noting that a rotation gesture of this manner only really needs to be concerned with the x and z axes; the y axis represents an unnatural twisting of the hand from left to right that has an extremely limited range of motion and is more likely to add extra rotation not desired by the user.

Let's define our render function and update it when we receive new hand tracking information (this time with a `Controller` to spice things up). We'll adjust the cube's rotation according to the palm normal:

```
        // start leap frame updates
        controller = new Leap.Controller();
        controller.loop(onLeapFrame);
    } // end of setup
```

Using a controller to receive frame updates has a couple of advantages over using `Leap.loop()`; you can ask the controller for past frames and you can specify some options to the constructor such as enabling gesture support and defining a frame callback method. Refer to the LeapJS API reference for more details.

We should probably define onLeapFrame:

```
function onLeapFrame(frame) {
    // don't bother unless there's a hand
    if (! frame.hands.length)
        return;
```

As seen earlier, we want to bail out early if we aren't planning on making any updates, to conserve CPU usage. While in this case it may not be a huge win, as there are rarely frame updates with no hands, it is good practice, especially since we are accessing the first array element right afterwards.

```
// get first palm normal
var hand = frame.hands[0];
var normal = hand.palmNormal;
  // use normal to rotate cube
cube.rotation.x = -normal[2] * Math.PI / 2;
  // normal z component
cube.rotation.z = normal[0] * Math.PI / 2;
  // normal x component
```

This may look a little puzzling, especially since we are using the normal's x coordinate to rotate the cube's z coordinate and vice-versa. Consider the fact that the normal vector values represent the palm axis rotation values, whereas the cube rotation values describe rotation around that axis. We throw in $\pi/2$ to limit the rotation to a quarter of a circle, which gives a nice level of sensitivity.

Now that we've rotated our cube, it is time to re-render our scene:

```
// update display
renderer.render(scene, camera);
```

Rotate your palm around and watch as the cube follows it, as if you were actually holding it in your hand. Take a minute to bask in the sublime glory of the future, right here in the palm of your hand.

Should you for some inexplicable reason decide to be difficult and end-userish, and actually resize your browser window while running this demonstration, you will end up with a rather stretched or squashed view, due to the render dimensions and camera aspect ratio not being updated to reflect the new window size. The following code will rectify that:

```
// keep aspect and render area consistent when the
window is resized
window.addEventListener("resize", function() {
var aspectRatio = window.innerWidth / window.innerHeight;
```

```
            camera.aspect = aspectRatio;
            // inform three.js that the camera properties have
    changed
            camera.updateProjectionMatrix();

            // should probably let the renderer know about the
    new
            // canvas dimensions too
            renderer.setSize(window.innerWidth, window.
            innerHeight);

            // now would probably be a good time to redraw
            renderer.render(scene, camera);
        });
```

After updating camera properties it is necessary to poke Three.js with a friendly reminder, updateProjectionMatrix(), to let it know that we've changed our mind about the initial parameters, and it should do us a big favor and reset its internal matrix. Don't call this method too often or it will get grumpy.

Summary

This should hopefully be enough of an introduction for using the controller on the web with 3D graphics to whet your appetite. This combination of technology has the potential to revolutionize everything from data visualization to gaming, and perhaps new applications not even conceived of yet.

Upon first learning the details of the controller, most people ask "what's it for?" There is no specific application; it was designed as a general input device to enhance the way in which users can interact with machines. There is no specific use case that the computer mouse was designed for, and despite the fact that its only use for a while was to enhance fancy word processing capabilities, the software to take advantage of the two-dimensional input evolved along with a general understanding by users of how to use it to manipulate software. If you have an interest in increasing the accessibility of computer-based services, getting creative and exploring novel ways of computer interaction, or just want to make something really cool, the Leap SDK is your ticket.

Index

W

WebGL 81, 82
WebSocket 72
window
  docking 58-61
WindowFlinger
  about 52
  implementing 52-55
Window management abstraction 55, 56
window.requestAnimationFrame() 87
window resizing
  handling 57
Windows Services For UNIX
        component 38

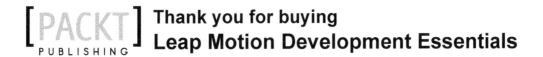
Thank you for buying
Leap Motion Development Essentials

About Packt Publishing

Packt, pronounced 'packed', published its first book "*Mastering phpMyAdmin for Effective MySQL Management*" in April 2004 and subsequently continued to specialize in publishing highly focused books on specific technologies and solutions.

Our books and publications share the experiences of your fellow IT professionals in adapting and customizing today's systems, applications, and frameworks. Our solution based books give you the knowledge and power to customize the software and technologies you're using to get the job done. Packt books are more specific and less general than the IT books you have seen in the past. Our unique business model allows us to bring you more focused information, giving you more of what you need to know, and less of what you don't.

Packt is a modern, yet unique publishing company, which focuses on producing quality, cutting-edge books for communities of developers, administrators, and newbies alike. For more information, please visit our website: www.packtpub.com.

Writing for Packt

We welcome all inquiries from people who are interested in authoring. Book proposals should be sent to author@packtpub.com. If your book idea is still at an early stage and you would like to discuss it first before writing a formal book proposal, contact us; one of our commissioning editors will get in touch with you.

We're not just looking for published authors; if you have strong technical skills but no writing experience, our experienced editors can help you develop a writing career, or simply get some additional reward for your expertise.

Kinect in Motion – Audio and
Visual Tracking by Example

A fast-paced, practical guide including examples, clear instructions, and details for building your own multimodal user interface

Clemente Giorio   Massimo Fascinari

Kinect in Motion – Audio and Visual Tracking by Example

ISBN: 978-1-84969-718-7   Paperback: 112 pages

A fast-paced, practical guide including examples, clear instructions, and details for building your own multimodal user interface

1. Step-by-step examples on how to master the essential features of Kinect technology

2. Fully functioning code samples ready to expand and adjust to your need

3. Compact and handy reference on how to adopt a multimodal user interface in your application

Kinect for Windows SDK
Programming Guide

Build motion-sensing applications with Microsoft's Kinect for Windows SDK quickly and easily

Abhijit Jana

Kinect for Windows SDK Programming Guide

ISBN: 978-1-84969-238-0   Paperback: 392 pages

Build motion-sensing applications with Microsoft's Kinect for Windows SDK quickly and easily

1. Building application using Kinect for Windows SDK

2. Covers the Kinect for Windows SDK v1.6

3. A detailed discussion of all the APIs involved and the explanations of their usage in detail

4. A practical, step-by-step tutorial to make learning easy for a beginner

Please check **www.PacktPub.com** for information on our titles

Cinema 4D Beginner's Guide

ISBN: 978-1-84969-214-4   Paperback: 274 pages

Model, animate, and render like a Pro!

1. Step-by-step instructions on modeling, texturing, lighting, and rendering a photorealistic 3D interior scene

2. Dynamic animations using MoGraph

3. Node-based programming to link parameters using XPresso

4. Stylized rendering with Sketch and Toon

Articulate Studio Cookbook

ISBN: 978-1-84969-308-0   Paperback: 292 pages

Create training courses with Articulate Studio's strong interactivity and rich content capabilities, all within the familiarity of Microsoft PowerPoint

1. Complete your courses by creating Flash-ready presentations through familiar PowerPoint

2. Employ Articulate Engage, Quizmaker, and Encoder to make dazzling interaction, assess learners, and add full-motion videos

3. Practical recipes to get you moving on a specific activity without the extra fluff

Please check **www.PacktPub.com** for information on our titles

www.ingramcontent.com/pod-product-compliance
Lightning Source LLC
LaVergne TN
LVHW080101070326
832902LV00014B/2362